WHY YOUR WORLD IS ABOUT TO GET A WHOLE LOT SMALLER

Jeff Rubin was the Chief Economist at CIBC World Markets for almost twenty years. He was one of the first economists to accurately predict soaring oil prices back in 2000 and is now one of the world's most sought-after energy experts. He lives in Toronto.

WHY YOUR WORLD IS ABOUT TO GET A WHOLE LOT SMALLER

OIL AND THE END OF GLOBALISATION

JEFF RUBIN

Published by Virgin Books 2010

2 4 6 8 10 9 7 5 3 1

Copyright © Jeff Rubin 2009

Published by arrangement with Random House Canada,
a division of Random House Canada Limited

Jeff Rubin has asserted his right under the Copyright, Designs
and Patents Act 1988 to be identified as the author of this work

First published in Great Britain in 2009 by
Virgin Books
Random House, 20 Vauxhall Bridge Road,
London SW1V 2SA

www.virginbooks.com
www.rbooks.co.uk

Addresses for companies within The Random House Group Limited can be found at:
www.randomhouse.co.uk/offices.htm

The Random House Group Limited Reg. No. 954009

A CIP catalogue record for this book
is available from the British Library

ISBN 9780753519639

The Random House Group Limited supports The Forest Stewardship Council
(FSC), the leading international forest certification organisation. All our titles that
are printed on Greenpeace approved FSC certified paper carry the FSC logo.
Our paper procurement policy can be found at www.rbooks.co.uk/environment

Printed and bound in Great Britain by
by CPI Bookmarque, Croydon CR0 4TD

TO DEBORAH, JACK AND MARGOT

[CONTENTS]

WHY YOUR WORLD IS ABOUT TO GET A WHOLE LOT SMALLER

REDEFINING RECOVERY

BEING AN ECONOMIST CAN RUIN YOUR APPETITE.

It is probably not the only job that has that effect. I've never worked as a taxidermist, but I can see that it might turn me off fish. My job, though, gets me worried about fish in a whole different way.

I like salmon—who doesn't? Salmon consumption has risen about 23 percent each year for the last decade or so. There are a number of good reasons to eat more fish: we all want food high in omega-3s, we want to eat less saturated fat, we want healthy protein for our low-carb diets. But here's the key reason for the amount of salmon on your dinner table: cheap oil has been subsidizing the cost of fish. Just like Wal-Mart and Tesco and big-box retailers around the world have been able to cut prices on almost everything by taking advantage of cheap shipping and cheap Asian labor, salmon went from being delicious local seafood to being another global commodity. Cheap oil gives us access to a pretty big world.

In the global economy, no one thinks about distance in miles—they think in dollars. If oil is cheap, it really doesn't matter how far a factory is from a showroom or a farmer's field from a supermarket. It's the cost of other things, like labor or tax, that determines what happens where. An Atlantic salmon caught off the coast of Norway is destined to be moved around the world just like a ball bearing or a microprocessor.

First the fish is taken to port in Norway, where it is frozen and transferred to another vessel, which will take it to a larger port, probably Hamburg or Rotterdam, where it will be transferred to *another* ship and schlepped to China—most likely Qingdao, on the Shandong Peninsula, China's fish-processing capital. There the whole salmon will be thawed and processed on a sprawling, neon-lit factory floor where squads of young women with nimble fingers skin, debone and fillet the fish. It will then be refrozen, packaged, stowed on another container ship and sent to a supermarket in Europe or North America. Two months after it was caught, the salmon will be thawed, displayed on crushed ice under gleaming halogen lamps and sold as "fresh."

Still, if I'm sitting in a nice restaurant and I'm enjoying a good conversation over a glass of wine, that is not what I am thinking about. And anyway, the shipping news doesn't normally appear next to a menu item. But if that conversation turns to energy and oil prices (and I confess it does fairly regularly), then when I glance at that fish I know I am looking at the past.

In the near future there is going to be less salmon on our tables—and probably fewer restaurants to eat in, too. Because the cheap-oil subsidy that makes Norwegian salmon affordable is about to disappear.

And as it does, your world is about to get smaller—much, much smaller.

To get that salmon from the ocean to your plate takes a ridiculous amount of energy. Think of the fuel for the fishing boats, container ships and just-in-time delivery trucks; the energy to freeze and process the fish, to sell it in a supermarket (retail stores use almost as much energy per square foot as factories do, just on heating, cooling and lighting). We invest a lot more energy to get that salmon than we get out of it when we eat it, which in itself makes the fish a bad energy deal. Economics calls it a "diminishing rate of return."

But it gets worse. A lot worse. All of that energy costs money, and energy gets more expensive just about every day. Not quite *every* day, of course—the recession that seemed to catch everyone by surprise in 2008 brought oil prices down in spectacular fashion. But even the deepest recessions last barely over a year. Those prices will be on their way back up soon enough. And however you want to measure the energy in that fish—calories, miles, joules, barrels of oil—it is inevitable that the price of fish is going to go up as well.

The seafood on your plate depends on cheap energy. And what is true of salmon is true of just about everything else. All you have to do to find an example is look around. Every morning when I head out to go to work, I see thousands of examples: the commuters making their way downtown from far-flung suburbs. The city I live in happens to be intersected by one of the busiest highways in North America—half a million cars make their way through its most heavily trafficked interchanges every day. Are those commuters going to be living or working where they are today when oil prices inevitably soar again? And if they are, will they still be driving cars? Either our living arrangements or our transportation options are going to have to change. In other words, our whole way of life depends on the price at the pumps, and that price depends on an uninterrupted supply of oil.

Think about that as you drive to work. Have a look at all those car dealerships, the gas stations and garages, the drive-thrus and big-box stores surrounded by huge parking lots. Try to imagine your life—picking up dry cleaning, taking your kids to hockey, going to Home Depot on the weekend, heading to the cottage in the summer—without a car. If you are like most people in North America or Australia, or even a less car-dependent country like the UK, you probably can't do it. And if you can't, you now have a small sense of what depends on the price of what comes out of the pump.

I say a small sense, because not only does your car burn energy, it is made from energy. Just building your car requires as much energy as it burns in several years. Add to that the fact that the plastics and paints and interior elements are made from petrochemicals derived from oil, and the picture becomes clearer. The house you live in is probably powered by electricity generated, at least in part, from hydrocarbons, and is almost certainly heated with natural gas or oil. The clothes you wear to work were probably made in some distant land and shipped here using relatively cheap oil, just as the coffee beans that went to make your latte were grown in a far-off country where the sun shines brighter and the labor is much cheaper, and then were shipped here.

So you see, it's not just your salmon. Despite the steady barrage of climate-change news and a growing sense that our affluent lifestyle may have unpleasant consequences for the environment, few of us stop to consider how just about every facet of our lives is built around our energy consumption. Nearly everything we do is inextricably bound to our use of energy.

And by "energy" I mean oil. Yes, we use natural gas and some coal to generate electricity; but the world's car and trucks and ships and planes run on oil. That means that the global economy runs on oil, because the global economy is about moving things around the world. And the reason the global economy has put all its eggs in one basket is that there is no other basket. As of right now, everything—from the salmon on your plate to the entire model of a global economy—depends on keeping the oil flowing.

Now, what happens when the price of salmon goes up? You buy less of it. And when the price of gasoline goes up, you drive less. When the price of clothes or computers or anything else goes up, everybody buys less.

And when everybody spends less, you have a recession.

It's not all that complicated. High energy prices cause recessions. A recession is not the end of the world, of course, though if you are one of the many people who has lost a job or seen your investments melt away, it can seem that way. Still, history keeps showing that the economy recovers, usually after a few quarters, and life goes on. Markets pick up, factories ramp up production, and eventually you're back to eating all the salmon you want.

But the history of the modern global economy is not all that long, and it is worth asking whether the patterns we have seen in past decades are ones we can expect to go on repeating into the future. We have seen high oil prices trigger recessions before, and in each case the medicine to cure a sick economy has been ready at hand: cheap new supply.

It's simple—as long as you have a ready supply of that oil.

But if you don't, the whole idea of recovery from a recession has to be redefined—because it's not going to look like it used to.

Right now, you need oil to make money and you need money to buy oil. If oil is too expensive, it becomes harder and harder to make money, whether you do that by driving a cab or by selling pineapples. And if there is no money to buy oil, the price of oil goes down. When it goes down, all of a sudden it's easier to make money again. But as long as you need oil to make money (and as chapter 7 will show, you do), the price of oil is going right back up once the money starts flowing again.

Sure, oil prices collapsed from record highs toward the end of 2008, but not before bringing down the global economy. It may be a record decline, but that says a lot more about where oil prices are coming from than it does about the price oil fell to. After all, oil prices have averaged over $40 per barrel since the recession was announced in the US in 2008. It wasn't that long ago that prices like that would have been considered pretty expensive.

But even more importantly, there is no way that oil prices are going to stay at these levels. As soon as the economy picks up, so will oil prices. That's because the fundamental causes behind triple-digit oil prices in 2008 won't have changed at all during the recession. In fact, they will likely have worsened.

As we will see in part 1 of this book, the reason the price of a barrel of oil hit record highs was that there is a deeply rooted imbalance between supply and demand. This doesn't mean speculators don't help push prices higher as well. Of course they do. But you have to ask why speculators got attracted to oil prices in the first place.

The answer is that they saw demand for oil rising relentlessly and they saw supply plateauing. It looked like a one-way bet, and in speculation, every day you're right, you've made money. Well, if you thought the price of oil could never fluctuate downward, you were wrong. Huge oil price increases have always caused recessions, so why wouldn't the recent record rise have the same effect? If you didn't see that, you probably lost some impressive sums of money. But if you figured there was a lot of demand chasing a relatively fixed amount of supply, you were right. And you still are.

What that means is that the moment the economy stops sputtering and comes back to life, oil prices will resume their upward trajectory. And all the sooner, since much of the new high-cost supply we were counting on in the near future has been canceled because of the decline in oil prices during the recession. The price of crude will keep going up until it triggers another downturn. As long as it takes a particular amount of oil to make a fixed amount of money or GDP, we are going to see our economies choked by rising prices almost as soon as they get back on their feet after each recession.

But it doesn't have to happen this way. One way to reduce the amount of oil we need to keep the economy running is to make your world smaller. And that is exactly what is going to happen.

—

I've got some good news and some bad news. Which do you want first?

That's what I should have asked the crowded dining room of oil executives in Calgary's Petroleum Club. As the chief economist at CIBC World Markets, a North American investment bank, I had come to talk about a subject very close to my audience's hearts: the future price of oil. I had something to say that should have improved their mood. But all they heard was the bad news.

The room was full of big personalities—anyone in the oil business had a plate full of the best Alberta prime rib money could buy. They were all there, from the big multinationals like Exxon, which owned much of the Canadian oil patch through its subsidiary Imperial Oil, to the small independents and aggressive entrepreneurs trying to make a living outfoxing the world's biggest oil companies. The one thing they all had in common that night was that one way or another they were all counting on producing more oil in the future, and they figured they were the guys who knew how it was going to happen.

Well, I figured they were wrong. I had just read an obscure but alarming study of the world's oil reserves called *The Coming Oil Crisis*, by Dr. Colin Campbell, a Cambridge-educated retired senior geologist who had spent the better part of his life exploring the world for new reserves. The title of the book pretty much gives away the ending.

What Campbell was suggesting was so contrary to the conventional wisdom about oil supply, and so staggering in its implications for the world economy, that I had decided to go see him. After a lifetime of oil exploration around the world, he had settled in a tiny Irish hamlet called Ballydehob, not far from Cork. I was in fact following in the footsteps of my wife, Deborah Lamb, who had recently led a Canadian Broadcasting Corporation film crew to do a documentary on oil depletion. There, in a tiny village along the Irish coastline, the world's most famous geologist explained to me what I was about to explain to Calgary's oil executives.

Campbell's argument was, and still is, that global oil production follows pretty much the same pattern as any individual oil well. Production at each reservoir accelerates until roughly half of the oil has been exhausted. Then it inexorably falls due to declining well pressure. A graph of oil production looks like a bell: a short, relatively horizontal line that steepens as it rises then flattens to a short peak before tracing a mirror image of its rise as it goes down the other side. The resulting curve—called the Hubbert curve, after the American geophysicist M. King Hubbert, who seems to have been the first person to figure out that there is only so much oil in the ground—gives us a pretty good visual impression of what we can expect from a finite resource: a peak, followed by a decline.

In 2002, Campbell first helped convene a loosely connected organization called the Association for the Study of Peak Oil to take an objective look at world oil supply. Pooling the experience of lifetimes in the field, the group of largely retired senior geologists who had explored the world for Shell, BP, Total, and all the other big majors built a massive database that tracked the depletion of every major producing oil field in the world. And when they added it all up, the composite picture that emerged about the growth of world oil supply was very different from the one their former employers were conveying. The rate of discovery was falling steadily and the rate of field depletion was rising just as inexorably. Run that model for very long, and pretty soon world production starts declining as well.

In other words, global production would soon be on the backside of the Hubbert curve. Campbell wasn't saying that the world was going to literally run out of oil. It never will, at least not in a time frame that matters to anyone reading this book. But daily world production, which had grown linearly until then, would soon peak, plateau and then begin its irreversible decline.

That struck me as a pretty important piece of news.

As an economist, I had been trained not to worry about resource limits—the question is not whether there is enough of something to go around, it's how much it will cost to get it *out* of the ground. And as someone whose job it is to forecast the economy, I knew how important oil—cheap oil—is to our economy's future health. In a word, *very* important. In fact, indispensable.

It was not long before I found that what economic theory was telling me was going to happen to prices was quite different from what Colin Campbell's anaylsis of depletion had to say. And if I was going to have to choose, I was going with the facts rather than the theory. The more I looked into the problem of oil depletion and scarcity, the more I found that looking at the problem as an economist usually told only half the story.

This book tells the other half.

—

But back at the Petroleum Club in Calgary in 2000, the Hubbert curve was going to be a tough sell.

You certainly were not going to hear about depletion from the oil companies. Their stock valuations depend in good part upon estimates of their reserve holdings. That makes "depletion" a dirty word in most oil-company boardrooms. OPEC producer states are even less inclined to talk about how quickly their countries' oil reserves are depleting. First, their production quotas are in part dependent on their reserve estimates. More importantly, since few will be capable of producing at their quotas anyway, candid depletion disclosure can expose a country to geopolitical as well as potential financial risk. That means that the only people who know exactly how much oil is in the ground are the last people who will ever tell you.

So I figured I would tell them. Oil had just reached a ten-year high of over $30 per barrel, after averaging around $20 per barrel over the last decade. Just about all of the oil and gas analysts out there, to say nothing of a similar percentage of economists, were predicting that OPEC would soon boost production and bring prices back down to their so-called target range. That is, most of the world, and certainly the folks I was speaking to that night, believed that we were in the midst of a temporary spike in oil prices that would soon be reversed.

I was ready to short that trade. I knew that the cartel had long ceased to be a price setter. They just didn't have enough production capacity to control prices any longer. They were a price taker, just like everyone else these days. If Campbell's supply projections were even remotely close to the mark, I knew that oil wouldn't be anywhere near the $20 per barrel range for very much longer. And when oil prices started to rise, they would have a long way to go up. I began modeling what oil prices would be like under increasingly restrictive supply conditions and came up with a forecast of $50 per barrel within five years.

So I took the stage to make the case that what we had just recently seen in oil markets was a harbinger of the future trend in world oil prices. Those high prices (remember when $30 oil seemed alarmingly expensive?) were not some cyclical blip or coincidence of special factors but the beginning of what would prove to be a spectacular rise in oil prices driven by a fundamental imbalance between ever-growing demand and ever-tightening supply conditions.

It is not just production that follows the bell-shaped curve toward the day when there is not enough oil to go around. Discovery of new oil fields peaked in 1966 and has been falling ever since. And while we still every once in a while read headlines about major new discoveries like the Tupi oil field off the coast of Brazil, announced to great fanfare in late 2007, what the oil companies don't hold glamorous press conferences to announce is that every year the world oil industry loses almost 4 million barrels per day in production through depletion. That is, as we drain the oil wells scattered around the globe, they produce less each year—a lot less. This means that the industry has to find roughly 20 million barrels per day of new production over the next five years simply to replace what will be lost. Right now, we are pumping about three times more than we're finding. That's a surefire recipe for even higher oil prices down the road.

And we almost never read press releases from oil companies or national governments explaining that what is coming out of the ground is not the cheap, free-flowing stuff that gushes out of the desert in Saudi Arabia, but the sticky tarlike bitumen that is mined from sands most of the year in subfreezing temperatures in northern Canada.

Not surprisingly, my supply forecast based on accelerating depletion of some of the world's workhorse oil fields and my forecast doubling of the price of a barrel of oil by mid-decade went over just like Hubbert's 1956 address to the American Petroleum Institute, where he first made the case that oil production in the United States would peak in the early 1970s and decline thereafter. He was laughed off stage, and his employer, Shell Oil, immediately disassociated itself from his forecast. Hubbert made himself an outcast by predicting the collapse of US oil production, and I was doing a pretty good job of wearing out my welcome too. My oil price forecast and its underlying supply projections were greeted with widespread amusement if not outright derision.

But as it turned out, Hubbert was right. American production peaked at just shy of 10 million barrels per day in 1971. It has fallen steadily since then. Today it is barely half that amount, at 5.1 million barrels. Tomorrow it will be even less.

And I was right too.

—

One thing I've learned from years of being on the opposite side of the peak oil debate from just about everyone else is that it is pretty much impossible to convince anyone of something they just don't want to believe. Campbell's forecast of a production peak was of course dismissed by the industry just as Hubbert's initial projections of a production peak were ignored decades earlier. Anyone who was willing to warn of pending supply shortages at a time of cheap and seemingly plentiful oil supply was ridiculed by the oil industry and consequently ignored by most of the media.

Still, I thought that might change when the facts started proving me right.

Five years after my first speech to the Petroleum Club, I returned to the same venue to give a forecast update on the outlook for oil prices. I was at this point feeling pretty confident of a more receptive audience, now that my earlier forecast was being borne out: the price of West Texas Intermediate, the benchmark North American oil (named after the sweet, light crude by which all oil refined in North America is judged), had already straddled my $50-per-barrel target. Maybe these folks weren't buying into the Hubbert curve, but a packed room told me a lot of people wanted to know what I was going to say next about world oil prices.

This time I was there to talk about demand, not supply. So far, the peakists had talked only about the threat to future oil prices from supply depletion. But depletion wasn't the only factor threatening future world oil supply. Explosive demand for massively subsidized oil in major oil-producing countries had become a new threat, particularly in the very places many of us are expecting to supply our future energy needs—the OPEC countries.

Feeling much more confident than five years earlier, and armed with new data on soaring domestic oil consumption in OPEC and other major oil-producing countries, I walked up to the podium and made the case that most of the world's major exporting countries were cannibalizing their own export capacity.

The price implications for world oil markets would be just as significant as those that followed from depletion. Unable to match strong demand growth in the developing world, increasingly restrictive conditions in world crude markets would send the oil price doubling to $100 per barrel within the next two years. In other words, the world's main oil producers would soon be burning so much of their own oil that there wouldn't be enough left for the rest of the world—like the United States, which burns about a quarter of the world's oil but pumps less than a tenth.

Well, I was wrong about one thing—a receptive audience. Anyone who has read the small print beside the asterisk in any mutual fund advertisement knows that past success is no guarantee of future performance. But I was still somewhat taken aback at how long people cling to past misperceptions. Despite my newly minted track record, no one in the room believed that oil prices were heading to $100 any more than they had believed years earlier that oil was going to $50.

Ironically, many of the very executives who snickered at my outlook turned out to be among the greatest beneficiaries of my forecasts—at least when they are proven right. Of course, oil shot well past $100, and the advent of triple-digit oil prices transformed the Canadian oil sands from a marginal resource propped up by huge royalty subsidies to one of the most important hydrocarbon deposits anywhere in the world. In the process, Calgary's Petroleum Club has been suddenly thrust from relative obscurity into the limelight of the world energy industry, triggering an enormous boom in Alberta, where for a while the people serving coffee in doughnut shops were making $40 an hour.

That was the good news I had tried to tell them five years earlier: high oil prices would suddenly make expensive tar-sands oil a hot commodity. But what is good news for Alberta is not necessarily good news for the rest of the world.

—

Every time the price of a barrel of oil dips by a few dollars, someone tells me I'm out to lunch. And when prices went from $147 to briefly below $40, not a few people figured I had been proven wrong pretty decisively. That's fine. I've had this debate on

CNN, on ABC, in the pages of the *New York Times* and the *Wall Street Journal*. There is *always* someone willing to argue that the tightness in world oil markets is due to a "perfect storm" of special circumstances that will soon pass and that once again all those silly notions of depletion will be proven wrong. When oil broke through the $100-a-barrel ceiling in January 2008, we were asked to believe that a single rogue trader had bid up the price to amuse himself on a day when many other traders were still on vacation. When it kept going up, other reasons were invoked, such as hedge funds piling into the market looking for easy money. One possibility we almost never heard about was the risk that oil supply might not meet demand.

I can't say I am surprised. After all, the conventional wisdom of economics says I should be wrong—supply *should* match demand.

The basic rules in economics are pretty simple, despite all the fancy mathematical packaging that comes with the discipline these days. The two fundamental axioms of the dismal science are that the demand curve slopes down and the supply curve slopes up. That is, the more people want something, the more it should cost. And the more it costs, the more of it there should be. Find the point of intersection between those two curves, and voilà, you have found the market clearing price.

If Porsche Carreras were given away to all ticket holders at NFL games, they would be worth a lot less than they are today. If we started running out of, say, shampoo, the price would go up. Manufacturers would have an incentive to ramp up shampoo production, and the price would come back down. Pretty simple.

The basic laws of demand and supply dictate that higher oil prices should draw additional supply from the ground while at the same time killing off demand. And that is exactly what economists keep predicting. Like Pavlov's dogs, that's what they are trained to do.

After all, they have history on their side, as the oilmen in Calgary were quick to remind me. Twice before, catastrophic spikes in the price of oil were followed pretty quickly by a return to normal prices, just as conventional economics would predict. In both 1973 and 1979, the world economy was thrown into chaos by fuel shortages and the high prices that accompanied them—only to see the time-tested laws of supply and demand quickly restore order to both oil prices and the economy at large.

And as economists predicted, higher oil prices triggered huge investments in technology that dramatically improved energy efficiency, whether it was smaller cars or natural-gas-fired electricity plants. It is also true that new supply was brought on line and helped force down the price of a barrel of oil. The British North Sea oil fields gushed oil into world markets, as did Prudhoe Bay in Alaska, helping restore global supply and fueling economic booms in the UK and Alaska. Once again the laws of demand and supply seemed to be working, with higher prices bringing forth the new supply that economics textbooks said they always fetch.

But history can be loaded with head fakes. The energy crises of 1973 and 1979 were political in nature, not geological or economic. The world began to run dry because major oil-producing nations simply turned off the taps. Eventually, they turned them back on again.

This time around, the tap is wide open. But even with everyone pumping as much oil as they can, what's flowing through the spigot these days doesn't seem to be enough to meet the world's growing thirst for oil. There is something far more fundamental going on.

In today's oil market, the laws of supply and demand have been turned on their heads. Contrary to the basic precepts of economic theory, global oil demand grew faster during the run-up in oil prices than it did a decade earlier, when prices were roughly a fifth or less of what they were in early 2008. Far from killing demand, record high oil prices seemed to spur ever-greater consumption of oil.

And instead of new supply gushing out of the ground, supply growth has basically stopped dead in its track in the face of no less than a fivefold increase in the price of crude. Despite every incentive to pump more, despite calls for OPEC to open the spigots and President Bush's personal pleas to the Saudis, world production has hardly grown at all since 2005.

Suddenly the textbooks seem to be describing some other world than the one we live in.

It is hard to say which possibility is more alarming to economists—that the world has reached its peak oil production plateau, or that the rules of their vocation don't seem to be working any more.

—

It is funny how a recession looks like good news to some people.

When global credit evaporated in the wake of the 2008 subprime mortgage crisis, oil prices tumbled along with the values of the world's stock markets. Seemingly overnight the price of a barrel of oil plunged from an all-time high of $147 to as low as the high $30s. Predictably, those who had piled into oil markets scrambled for the exit doors, especially hedge funds and other investors who were forced to sell their oil positions to come up with some money to cover the losses they were sustaining in the rest of their portfolios. And, just as predictably, what many observers concluded from watching prices fall was that there must not have been an energy scarcity problem after all, and that triple-digit prices had been just a speculative blip.

Of course, most of the commentators saying that were people who had never thought oil prices would ever get above $50 per barrel in the first place. Sure, if you think the market is going to solve the problem of high oil prices and then the price drops, you might be tempted to think that the market has done what you had such faith it would.

But no one said that oil prices will never fall. In fact, increasingly wild and destructive movement in prices is exactly what you would expect in an environment of global scarcity. Oil demand will drop in a recession, and so will the price of oil. So that can't be a surprise to anyone.

But we shouldn't be looking at oil prices as the effect of the recession. *They are the cause.* While the financial crisis from the imploding US subprime mortgage market gets top billing for the 2008 recession, the ascent of oil prices to record triple-digit levels played a far more major role in derailing growth in the North American and European economies.

To claim that the price decline is evidence that record prices were the consequence of massive speculation in oil markets is to ignore the underlying problem: a fundamental mismatch between global supply and demand. But what today's skeptics don't explain is why oil prices aren't $20 per barrel, as they were only eight years ago, during the last recession. West Texas prices have hovered around $40 per barrel, and Brent prices, the European benchmark, have traded around $45 even though this recession is well over three times as severe.

There is a good reason prices won't fall that far. The skeptics may not want to talk about it, but at $60 to $90 per barrel, many of the world's largest energy megaprojects, such as the Canadian oil sands, won't go ahead because those prices will no longer provide a sufficient economic return. Finding pocket change is getting pretty expensive these days and it's not going to get any cheaper tomorrow. If you believe that high prices bring new supply out of the ground, you are pretty much committed to the fact that every drop in price means that there is less oil to go around. There may be oil out there under the ground, but no one is going to sign up to lose money pumping it. The laws of economics cut both ways.

In any case, as we will see, it matters less every day how much oil is consumed by the countries of the Organization for Economic Co-operation and Development (OECD), a club of the world's thirty most advanced and wealthiest democracies. We may be easing off on demand in North America and Europe, but elsewhere in the world drivers and policymakers are getting on the accelerator even more enthusiastically than we are getting off. We can cut back as much as we like, yet as long as the Saudis and Venezuelans, the Chinese and Indians keep their feet on the gas, it is not going to matter.

In August 2008, when oil prices peaked, Americans drove 15 billion miles fewer than the previous August, the largest drop since the government started collecting data in 1942. That kind of collapse in demand is part of the reason for the decline in prices. But there are plenty of drivers elsewhere in the world who are more than happy to drive those miles and burn that oil. Even if demand were to stagnate in the rich countries, it is only going to grow elsewhere and eventually catch up to where we were when prices were so high.

But demand is not going to stagnate forever. This recession may be the deepest post-war downturn, but that is just testament to the destructive power of triple-digit oil prices. If $40 is as cheap as oil gets in the most severe recession, what happens to oil prices when the economy picks up again?

Simple. Once the dust settles from the various crises rocking financial markets, we are looking at the same basic demand–supply imbalance that we were looking at before the recession began. That imbalance took us to nearly $150 per barrel before the recession set in. In the next cycle, the same imbalance will probably take us to $200 per barrel before another recession temporarily knocks back prices and demand.

Economic activity goes hand in hand with energy use. If you want to grow the economy, you need to burn more energy — that's precisely why dwindling oil reserves pose such a threat to global economic growth. If instead the economy falters and begins to contract, less energy is used and hence its price will fall. That doesn't mean that triple-digit oil prices were a temporary aberration, but it does give a sense of how hard it is to keep the world economy running on cheap oil and it should make it pretty clear what happens to oil prices when the recession is over.

Other than lulling us into an unjustified sense of optimism about the future direction of oil prices, a global recession will do absolutely nothing about the unavoidable fact that oil production is nearing a plateau while oil consumption around the world is still rising. Recessions don't diminish our dependence on oil; they just cut back a little on our appetite for it. When we start to feel a little better, we will be guzzling it again, and we may well be left wanting more. Because unlike after past oil shocks, there is no post-shock boost in oil supply to look forward to any more.

If we wait for Adam Smith's invisible hand to pull abundant sources of new cheap oil out of the ground, we are going to be waiting for Godot. Governments around the world may be thrusting bailout money into the hands of businesses and taxpayers, but you can count on one thing. There will be no energy bailout.

—

Just as I had good news and bad news for the oilmen, part 2 of this book will have good news and bad news for you too.

First the bad news. With supply dwindling and demand rising, you can expect scarcity. And scarcity means high prices. You can expect triple-digit oil prices in the near future. Yes, the price at the pump is going to go up. Count on it. In the US, that should translate into as much as $7-per-gallon gasoline, and about $2 per liter in Canada. Europe is of course already paying those prices, so they should get ready for the equivalent of double-digit gas prices. But it will also hurt in a lot of ways you may not be thinking about.

Life as we've known it is up for grabs in a world of expensive fossil fuels. Expensive oil means a severe curb on the free-spending lifestyle that cheap energy has afforded us for some time now. It means you can say a long and wistful goodbye to the inexpensive products manufactured on the other side of the world. You may not love them, but they have been stretching our dollars for a while now and holding down inflation at the same time. You'll miss them when it starts to become clear that your paycheck just doesn't go as far as it used to.

Your food in particular is going to cost a lot more—in fact, it is already getting more expensive all the time. The stuff you burn in your car is the same thing the farmer in Iowa needs to plant and harvest his corn (to say nothing of the natural gas needed to manufacture his fertilizer). It's the same stuff that powers all the trucks and planes and ships that move everything around, the same stuff that is used as a feedstock for the petrochemical industry that produces our plastics and pharmaceuticals. It's what the navy uses to fuel its ships, and what the local government needs to run its lawnmowers to keep the parks looking groomed. Someone is going to have to pay for all of this, and less oil means less money. Some difficult choices lie ahead.

Now the good news.

Expensive oil may mean the end of life as we know it, but maybe that life wasn't particularly great to start with. Smog-congested cities, global warming, oil slicks and other forms of environmental degradation are all part of the legacy of cheap oil. If you want a hint of what the future will look like if oil-guzzling members of the OECD get it right, just look at Europe today. There, drivers are already paying the equivalent of $7 for a gallon of gasoline, and in France and Germany life goes on.

European gasoline prices give a hint of what is down the road for North Americans, and it is not all doom and gloom. Sure, we will be facing higher prices (if you've ever bought a pint of beer in Frankfurt or a latte in London, you know just how much higher European prices can be than what North Americans pay). We will be living in denser communities, driving smaller cars, living more frugally and locally. When we travel, we may soon be boarding an electric-powered train rather than an oil-powered airplane. And with global climate change also bearing down on our energy consumption, we may be soon be paying more attention to the cost not only of buying carbon-based fuel, but of burning it too, just as the Europeans are already doing.

But living in a clean, efficient, densely populated city is not exactly the end of the world. Where would you rather spend your vacation: Paris or Houston?

And while there are certainly going to be losers as the

eighteen-wheeler of globalization is thrown into reverse, there are going to be winners too. In a world of triple-digit oil prices, distance suddenly costs money and lots of it. Many of those once high-paying manufacturing jobs that we thought we had lost forever to cheap labor markets overseas may be soon coming back home. With every dollar increase in the price of the bunker fuel that powers the container ships that ply the Pacific, China's wage advantage becomes less and less important and Western workers once again become competitive. Who would have dreamt that triple-digit oil prices would breathe new life into America's rust belt or the British steel industry?

Get ready for a smaller world. Soon, your food is going to come from a field much closer to home, and the things you buy will probably come from a factory down the road rather than one on the other side of the world. You will almost certainly drive less and walk more, and that means you will be shopping and working closer to home. Your neighbors and your neighborhood are about to get a lot more important in the smaller world of the none-too-distant future.

Here's the question: will we decide to reinvest in a global economy and an infrastructure that keeps us bound to oil consumption for every dollar or pound or yen of wealth we produce? If so, we are committing ourselves to a damaging cycle of recessions and recoveries that keeps repeating itself as the economy keeps banging its head on oil prices. If we go this route, peak oil will soon lead to peak GDP.

Or we can change. Not only must we decouple our economy from oil but we must reengineer our lives to adapt to a world of growing energy scarcity. And that means learning to live using less energy. While much could go terribly wrong in this transition, don't be surprised if we find more than a few silver linings in the process, like a solution to carbon emissions for example. And don't be surprised if the new smaller world that emerges isn't a lot more livable and enjoyable than the one we are about to leave behind.

Either way, your world is about to get a lot smaller.

PART ONE

CHAPTER 1

SUPPLY SHIFT

WHEN YOU DON'T HAVE ENOUGH MONEY in your pocket to get on the subway, finding some change between the cushions of your sofa can seem like a pretty important discovery. And if it gets you to work on time, it is.

But if you had to depend on finding spare change to get to work each and every day, well, you might start looking for another job a lot closer to home.

Right now, the oil companies of the world have their hands deep between the cushions, and so far they have been coming up with enough dimes and quarters (or pence, for that matter) to get us all to work. But there is only so much change to be found, and more people are heading out the door to work every day.

The world's oil wells are running out of the stuff that keeps the whole system going. Every well will eventually run dry—in fact, a well is tapped out long before it is empty. Even with the most advanced technology, oil companies get barely half of the oil out before geophysics and economics conspire to make what's left not worth going after. The question is how fast, and whether new supply can be found in time to replace it. Even if today's depletion rate holds constant, we must find nearly 20 million barrels per day of new production over the next five years simply to keep global production at its current level. That's a tall order even if you are talking about easy-flowing, cheaply produced conventional crude oil. But that's not what we are talking about at all.

What we are talking about instead is the energy equivalent of pocket change. The oil that is coming on line to replace the declining wells is dirty and hard to find. It is unreliable and expensive. Like the change you desperately look for in the pockets of pants you haven't worn for a while, a lot of the oil we refine and burn these days is stuff we wouldn't even bother with if we still had access to yesterday's oil fields.

But we don't have access to those fields any more. They have long been sucked dry by our parents' cars. From the North Sea to Mexico and from Russia to Indonesia, oil wells are producing less oil than they used to. For a long time, the International Energy Agency (a body set up by the OECD to prevent supply disruptions and promote energy security) kept telling us the world's wells were declining at 3.7 percent each year. Then, in late 2008, after a rigorous international study, they revised that figure to 6.7 percent. At the stroke of a pen, the global economy lost millions of barrels of oil.

The situation is particularly dire in the world's biggest oil market—the United States. Over the last three decades, US oil production has literally halved, from as much as 10 million barrels per day in the early 1970s down to 5 million today. Over the last 10 years alone it has dropped by half a million barrels per day in the lower forty-eight states.

Meanwhile, in Alaska, which not long ago was thought of as a new energy frontier, production has been declining even more rapidly than in the rest of the country. It was in part Alaskan oil from Prudhoe Bay that helped break the OPEC stranglehold that provoked the energy and financial crises of the 1970s. Prudhoe Bay is still the biggest oil field in North America — more than twice the size of anything in Texas. But Prudhoe Bay oil was discovered in the late 1960s and came on line in 1977. Since then, Alaskan oil production there has fallen by almost a million barrels per day.

That's fine if you're burning less oil. But Americans burn more oil than they did three decades ago. Oil consumption in the United States has gone from 15 million barrels per day in 1970, back when American oil wells were still fueling American cars and trucks, to almost 20 million today. Needless to say, the difference between what is produced in America and what is burned in America is made up by imports, with all the trade imbalances and geopolitical difficulties that go along with depending on foreign suppliers for the one thing necessary to run the economy.

The story is the same whether you live in the US or the UK or anywhere else in the developed world. In fact, while consumption has grown by about 20 percent in Canada and the US since 1980, it is up 63 and 74 percent respectively in Australia and New Zealand. When consumption is up and supply is down, it is hard not to worry about tomorrow's oil supply, and the implications for tomorrow's economy as well.

Still, in 2004, the US Department of Energy announced that good news was on the horizon. Growing oil production in the Gulf of Mexico would be America's great hope for an energy resurgence, delivering the country not only from oil depletion but from the political clutches of hostile supplier nations. They confidently predicted that production in the Gulf would double to 2 million barrels per day by the end of the decade and then double again to 4 million barrels per day by 2020.

A whole lot is at stake in the efforts to boost production in the United States by drilling offshore. The Gulf of Mexico is the only place where production has grown at all over the last fifteen to twenty years. It has been the single bright spot against an increasingly gloomy background of falling domestic production.

The energy bureaucrats in Washington, along with the oil industry, were certainly bullish. They ridiculed predictions of depletion of American oil supply, claiming that such forecasts ignored the vast treasure chest of oil and gas reserves that lay trapped under the floor of the Gulf of Mexico. And good old-fashioned American know-how and technology was going to bring it to your local gas station in no time at all.

Work was proceeding at a feverish pace on British Petroleum's Thunder Horse platform over the spring of 2005, the centerpiece of the huge planned expansion of underwater oil production in the Gulf of Mexico. Adorning the cover of *BusinessWeek*, Thunder Horse defined the new frontier of US oil supply and the new frontier of underwater oil technology.

To get its hands on all this oil, British Petroleum had to drill 120 miles off the Louisana coast, in water over a mile deep—and more than 3 miles below the ocean's floor. The actual field area is over 50 square miles, its oil pumped through a floating platform nearly the size of a football field.

Thunder Horse was expected to produce as much as 250,000 barrels per day when it opened later that year. But the idea was that it could be expanded to produce over half a million barrels per day. Even at its initial flow rate, it would be the largest producing field in the US section of the Gulf of Mexico.

At least that was the game plan. But Thunder Horse— and indeed the entire Gulf of Mexico oil industry—was caught flat-footed when the hurricane season arrived that year and the once-placid waters of the Gulf began to stir. The season started off with a bang, setting a record for the earliest recorded storm. And it built from there with no less than a record four Category 5 storms. But it was Katrina and Rita, more than any of the other eighteen hurricanes in the Gulf and the Caribbean that year, that shocked the world—particularly Katrina.

The oil and gas industry's drilling platforms that dot the Gulf didn't fare much better than the New Orleans levees against Katrina's storm surge. The design standards the industry proudly touted as up to "storm of the century" standards proved more than a tad less durable when confronted with the real thing.

Together, Katrina and Rita ravaged no less than 167 offshore platforms and 183 pipelines. Many platforms were tossed about like inflatable dinghies in the waves, with uprooted remnants found as far as 60 miles away from their original moorings. It turned out that few if any drilling platforms could withstand the force of a Category 5 hurricane and most couldn't even weather Category 3 or 4 storms. As for the mighty Thunder Horse, it was toppled the month before Katrina when Hurricane Dennis left the billion-dollar rig leaning drunkenly in the waves.

At the time, the oil and gas industry argued that both hurricanes Katrina and Rita were freak events unlikely to be repeated. But only three years later they were proved wrong when hurricanes Ike and Gustav came smashing through the Gulf. But even before Katrina and Rita struck, there was already compelling evidence that hurricanes had become much more destructive in the region of late.

Whereas severe storms (Category 3 to 5 hurricanes on the Saffir-Simpson scale) had once been relatively infrequent events, they are becoming increasingly common in the Gulf of Mexico. Indeed, the National Oceanic and Atmospheric Administration (NOOA), the official US weather agency, has found that over the last three decades, the proportion of storms that falls into these categories has doubled. Moreover, the NOAA expects that the trend is likely to intensify over time as the earth's climate is warmed by increasing levels of greenhouse gas (GHG) emissions in the atmosphere.

Ironically, the United States, a holdout on signing the Kyoto Protocol on greenhouse gas emissions, may soon find that its densely populated and energy-rich Gulf region has become part of the world's early warning system on global warming.

Warm water is essentially hurricane fuel. The increasing frequency of severe tropical storms in the Gulf of Mexico has been found to be strongly correlated with the rise in sea surface temperatures in the Gulf and in tropical latitudes of the Atlantic Ocean where most Gulf hurricanes originate. That means that what was once a storm of the century will now be the storm of next year.

It wasn't oil prices that North Americans remember most about the impact of the 2005 hurricane season, but the prices at the pump. While the price of West Texas Intermediate rose by roughly 10 percent, to $70 per barrel, the price at the gas station rose by 50 percent, to $3 per gallon. Even 1,000 miles north of sidelined Gulf Coast refineries, gasoline prices in Toronto shot up from under $1 per liter to about $1.30, mimicking price movements south of the border. And the reason wasn't just the storm-induced rise in oil prices. It was also the storm-induced rise in crack spreads.

It costs money to turn oil into gasoline, a process called "cracking." Cracking involves heating crude oil and adding hydrogen in order to turn it into a lighter, more valuable fuel like gasoline. The cost difference between what goes into a refinery (oil) and what comes out (gasoline) is called the "crack spread."

Prior to the 2005 storm season, crack spreads in the Gulf of Mexico were averaging around $10 per barrel. That means essentially that the outgoing barrel of refined gasoline cost $10 more than the incoming barrel of crude oil. The trouble is that 40 percent of US refinery capacity is located along the Gulf of Mexico coastline, with a particular concentration around the Houston area. As the hurricanes flooded refineries and knocked them out of production, there was suddenly not enough capacity to process the fuel to keep the gas stations pumping. Gasoline shortages quickly ensued, sending pump prices skyrocketing. Crack spreads went through the roof. The more oil there is chasing refinery capacity, the wider the crack spread. In this case, the laws of supply and demand work like a well-oiled machine.

There are 42 gallons of oil in a standard barrel. When refining costs exploded from $10 per barrel to over $50 per barrel in 2005, that added 75 cents to the price of a gallon of gasoline at the pumps. And that's before taking into account any increase in the price of oil itself. All of a sudden American motorists were paying the gasoline price equivalent of nearly $100-per-barrel oil even when the actual cost of oil was a good $30 per barrel less.

That story replayed more or less in September 2008, when crack spreads soared even higher as Hurricane Ike laid a direct hit on Houston and the heart of America's oil-refining industry. While oil prices didn't move like they did when Katrina and Rita hit, gasoline prices soared to as much as $5 per gallon in many Gulf states.

The damage to American refineries and the attendant spike in gasoline prices underscores the double jeopardy facing American motorists. Not only do hurricanes threaten roughly a quarter of US oil production in the Gulf of Mexico, just as importantly they also threaten about 40 percent of US refinery capacity—and hence the ability to process alternative sources of crude supply. That is, even if the oil companies could get their crude from somewhere else, they still wouldn't be able to refine it into gasoline.

What does all this mean for the optimistic forecasts that Gulf production would reverse or at least stem the decline in American oil supply? The US Department of the Interior's Minerals and Management Service estimates that as of the summer of 2008, three years after the storms hit, Gulf production was still 20 percent, or nearly 250,000 barrels per day, lower than it was before Katrina and Rita hit the region. Repairing the damage after the 2008 storm season will probably see production in 2009 fall even lower. But that measures only the physical damage.

Storms don't just take away today's oil. They take away tomorrow's too. Instead of bringing new fields on line, the oil and gas industry spent three years repairing badly damaged fields like Shell's giant Mars platform. Hurricane-proofing old rigs means not spending the time and money required to build new ones.

This is not a matter of putting off until next year what you would otherwise have done this year. When you face double-digit-percent annual depletion rates, as you do in most underwater fields in the Gulf of Mexico, you have to constantly be bringing on new fields just to keep production from slipping; if you want production to actually grow, you need still more new wells. And if you expect to double production, as the US Department of Energy claimed would happen in the region, you're going to need a very ambitious building and drilling program.

With 2008's damage from Gustav and Ivan causing a brand-new round of shut-in production and project delays, US oil production in the Gulf of Mexico is heading backwards, not forwards. While hurricanes continue to postpone new production, depletion doesn't miss a beat. More than three years after Katrina and Rita hit, not only is Gulf production miles below the optimistic forecasts of the US Department of Energy, the region actually produces a quarter of a million barrels per day less than it did only half a decade earlier.

Maybe the American oil industry is finally figuring out why the Mayans built their major cities inland. When you live in a region for a millennium or so, you get to know the weather.

RUNNING FASTER TO STAND STILL

The oil fields of the Gulf of Mexico have one more fatal flaw, which they share with other deep-sea projects around the world. Underwater fields, especially deepwater fields, deplete twice as rapidly as conventional oil fields on land.

Depletion rates don't increase despite technological advances — they increase *because* of these advances. New technology doesn't put any more oil in the ground. It just means we have a bigger straw to suck out what is already there.

Deep-sea oil rigs are big straws, in large part because deepwater drilling technology was developed later than conventional techniques. The North Sea flooded the world with oil and Great Britain with petro-wealth when the giant Forties and Brent oil fields came on line in 1975. But only a couple of decades later, production rates were dropping like a rock, no matter how much money and technology were thrown at them. In fact, by the 1980s the oil companies had already spent more on drilling technology in the North Sea than NASA spent putting a man on the moon. But the North Sea wells had their best monthly production numbers in 1985. Their peak annual production came in 1999, and had dropped by 43 percent by 2007. While not long ago the

UK was an oil exporter to be reckoned with (indeed, the benchmark North Sea oil is Brent Light Crude), today it is a net importer.

And that process is repeating itself worldwide. The peak year for deepwater oil discovery was back in 1996. Given past lags between discovery and peak production for other sources of oil supply, a peak in offshore production is not far away.

Why does it matter if offshore production will soon peak? Because since 2000, offshore fields, and particularly deepwater fields, have been the single largest source of new supply growth in the world. But they are also the key reason behind an accelerating global depletion rate. The more dependent we are on deep-sea oil, the more depletion is going to accelerate.

New discoveries are the antidote to depletion. But while discoveries are constantly being made, today's new oil finds are a shadow of yesterday's. Take the much-touted Tupi field off the coast in Brazil, for example, and compare that to the type of fields the industry used to discover. In 1901, the Spindletop gusher, near Beaumont, Texas, shot oil 150 feet into the air and tripled US oil production overnight. Ghawar, in Saudi Arabia, was discovered in 1948 and is still the most productive field in the world. While Spindletop was barely 1,100 feet below ground, Tupi's oil lies beneath more than 2,000 feet of water, 10,000 feet of rock, and 6,600 feet of salt—not the kind of easy-flowing, bubbling crude that Jed Clampett of the *Beverly Hillbillies* found while shooting possum in east Texas. Tupi sounds pretty promising at 8 billion barrels until you take into account that Ghawar once held 60 billion barrels. Even Tupi starts to look a little like pocket change.

The rapidly changing nature of our oil supply is something that economists have a real hard time coming to grips with. That's primarily because from an economist's perspective, natural resources are effectively limitless. Even if resources become scarcer, economists have always argued, the higher prices that come with scarcity will encourage us to look for more efficient extraction technologies to get to harder-to-find deposits. That is, when oil costs more, there is more of it. A barrel of oil that costs $100 to extract is no good to anybody when the world price is $20. It's going to stay in the ground. But when the price goes up, suddenly that expensive oil becomes worth going after.

The cherished principle that higher prices will always raise more supply typically blinds most economists to the growing geological challenge of finding new oil. More than any other single precept of economics, the theory of the upward-sloping supply curve is the major reason why economists as a profession have been so dismissive of any notion of resource depletion. The only question that resources pose for conventional economics is the cost at which we can consume them. Physical supply, relative to the scale of potential human consumption, is treated as if it were infinite. And nowhere has the notion of infinite supply been more firmly etched into the economist's mind than when it comes to oil.

But where is the big increase in oil supply that economists' theory of the upward-sloping supply curve keeps predicting? For nearly a decade, people like Daniel Yergin at Cambridge Energy Research (CERA) have been assuring us that rising prices would bring forth more and more oil. But despite the confident projections of economists and energy-industry prognosticators, only a trickle of new supply has flowed, and a very expensive and problematic trickle at that. Even the oil industry is beginning to acknowledge that if we are not staring at a production peak in world oil, we are at a minimum entering a period of unprecedented scarcity.

That may come as a surprise to many in view of the regular stream of announcements of huge supply increases from the various oil megaprojects around the world. The problem is what you don't hear about. What oil companies and oil-producing countries don't send out splashy press releases to announce is how much their wells are losing to depletion every year. Each year, less and less flows out of the world's mature oil fields, particularly the relatively low-cost fields in the Middle East. These fields have been the backbone of world supply since the years following the Second World War, when Saudi Arabia started to dominate the world oil market with its staggering abundance of free-flowing oil. Both the world's largest conventional oil field (Ghawar) and its largest offshore field (Safaniya) are in Saudi Arabia. Today, though only the Saudis know for sure, there is every reason to believe both are depleting. Life without Ghawar and Safaniya would mean a whole new normal for the global oil market.

Depletion means you have to run faster to stand still. And the treadmill seems to be creaking faster and faster all the time. Depletion from existing fields has accelerated to a rate that now takes roughly 4 million barrels per day from world production every year. That's a tall order to replace. It's become virtually impossible to both replace and at the same time add to new global production in order to keep up with a world that every year wants to consume more oil.

And now that the global depletion rate has been rising alongside our growing dependency on deepwater oil, it gets even harder to stand still. Over the next five years we will need to replace nearly 20 million barrels per day of cheaply produced conventional oil with extremely problematic and high-cost unconventional supply—because that is the only new supply that higher oil prices can raise any more.

Between 2005 and 2007, even with record oil prices of recent years bringing new high-cost sources of supply like Canadian oil sands to the market, world oil supply did not effectively grow. If those prices didn't pull new supply out of the ground, what are the chances that prices falling in the wake of a deep global recession are going to do the job today?

What is coming out of the ground and trickling into world markets—pretty much all of the reported increase in world oil supply between 2005 and 2008—are fuels that you can't burn in your car. Included in the International Energy Agency's (IEA) estimates of "oil" are natural gas liquids such as butane, which commonly occur alongside oil production. Though these are valuable hydrocarbons in their own right, they are not an economically viable feedstock for either gasoline or diesel, the world's two most sought after transport fuels—pretty much the *only* transport fuels.

Imagine running out of gas on a lonely highway late at night and waiting for a passing driver to flag down. When a car finally stops, out jumps a friendly stranger with a canister of butane. You might be grateful for the thoughtful gesture, and you may be able to refill your cigarette lighter, but neither you nor your car is going anywhere.

Whether this proves to be a temporary aberration or a harbinger of future supply conditions remains to be seen. But many have noted that the increase in natural gas liquids found from aging oil fields is a telltale sign of accelerating depletion.

Not surprisingly, natural gas liquids are typically found with natural gas. As an oil field matures from cumulative resource extraction, the resulting loss of reservoir pressure releases trapped natural gas that had been suspended in oil. Over time, the released natural gas forms an expanded cap over the oil field, resulting in a rising ratio of natural gas to oil and in a rising ratio of natural gas liquids to oil. For that reason, the relative growth in natural gas liquids is suggestive of accelerating depletion in oil fields.

So not only are we not getting what we want, but what we are getting looks a lot like geological proof that there is not much left of the stuff we want.

Whether we are already at a production peak or not will be evident only in hindsight. But the precise date is not really the point. It is already clear that we are in the midst of a quantum supply shift from relatively low-cost conventional oil to high-cost and highly problematic nonconventional oil. And the one common denominator found in all forms of nonconventional oil is that it is very expensive oil, for no other reason than the fact that it takes a lot of energy to get it out of the ground.

As we come to rely more on increasingly costly nonconventional oil, we can only expect the price of oil to rise accordingly, even if future production targets are met. That is, simply meeting supply targets is not going to keep the price of oil down if the cost of producing that oil is higher than what it costs now. But when it comes to bringing some 20 million barrels per day of new oil production on stream from places like Sakahlin-II in eastern Siberia, from technically challenging fields like the Kashagan project in Kazakhstan, or from oil sands in the frozen muskeg of northern Canada, production never goes according to plan.

Only a decade ago, the Caspian basin was being hailed by the oil industry as the new Middle East. (Never mind that it is also the old Middle East. At the turn of the last century, Baku, Azerbaijan, was the oil capital of the world and boasted 3,000 wells.) And no field held more promise than the Kashagan field. Some claimed it would be the next Ghawar. But Kashagan, like so many other white elephants the industry has flogged, has turned out to be a nightmare. There is so much deadly hydrogen sulfide in the oil that workers must carry gas masks on site as a precaution against asphyxiation. Plagued by endless cost overruns, first flow dates have been pushed from an original 2008 to beyond 2013.

Welcome to the new frontiers of world oil supply. They are in some of the most remote and inhospitable places on the planet. If there has been one constant the world can safely bet on when it comes to bringing oil to market from places like this, it is that it takes three times as long and costs three times as much as the operator or host government claims it will. Shifting from conventional to nonconventional oil means that delays and mammoth cost overruns will become the norm of, not the exception to, tomorrow's supply picture. In oil, like in any other resource-extraction industry, you don't leave the easiest for the last.

And like on any new frontier, there have been more than a few shootouts over the spoils. As development costs soar and become multiples of original estimates, tensions have risen between companies and host countries, prompting major changes in royalty agreements or even changes in ownership. Just ask the executives at Shell about their former Sakahlin-II project, one of the biggest they had going. The company sunk billions of development dollars into this series of offshore rigs in the frigid waters of the Okhotsk Sea in the North Pacific. When staggering cost overruns were about to cut the Russian government out of its royalty share, the project all of a sudden ran afoul of previously nonexistent Russian environmental regulations, and Shell found itself an unwelcome guest. Under duress, Shell was forced to sell out to Russian interests and walk away from 1.2 billion barrels of oil.

If Shell shareholders were asking their board why the company risked billions of dollars in a both geologically and politically challenging environment in eastern Siberia, the answer is simple. That's all that's left.

TIME OF SANDS

Nothing illustrates that fact better than the Canadian oil sands. Like the recent growth in the production of natural gas liquids, the fact that so many billions have already been invested in the oil sands is a pretty strong sign that there is not a lot left of the light, sweet, easily pumped crude the oil companies and their customers actually want.

If there were, they would be pumping it, not mucking around in northern Alberta, where temperatures drop to –20 degrees Fahrenheit in winter and each barrel of oil is soaked into two tons of sand. To get it out first requires a fleet of multi-million-dollar dump trucks capable of carrying more than three hundred tons of cold, oily sand. (In case you're wondering, replacing a flat tire on one of these 24-cylinder, 1.3-million-pound brutes is going to cost you about $35,000.) You're also going to need a number of colossal steam shovels, each of which burns through 4,250 gallons of diesel each day running around the clock to dig all this stuff up.

Perhaps most importantly, you're also going to need a compliant host government to allow you to wring profits from the devastated landscape.

The attraction of the Canadian oil sands isn't just about what's under the ground—although 165 billion barrels of oil is not something an oil-hungry world is likely to ignore—but, just as importantly, where it is located. Canada is one of the few remaining places in the world where private companies can own oil and develop oil resources. While most North Americans might take for granted the right of an Exxon to come in and profitably exploit an oil field, that's certainly not par for the course in the rest of the world these days.

As in real estate, location is everything in the global energy industry. Depending upon your view of the current investment climate in Kazakhstan and Nigeria, the Canadian oil sands represents anywhere from 50 to 70 percent of the investable oil reserves in the world. By "investable" I mean that privately owned (and largely foreign) oil companies can obtain from the host government concessions to own and develop hydrocarbon resources on their territory.

Even more importantly, it means that after spending billions of dollars on developing an oil property, oil companies don't have to worry that the government is going to suddenly change its mind and take that property away, like Russia did to Shell and Venezuela did to Exxon's properties in the Orinoco oil sands. Where are you going to invest—in a country that just seized billions of dollars worth of your assets (or even your competitor's), or in the Great White North, where sponsoring a minor hockey team in Fort McMurray is about all a multinational oil company has to do to be a good corporate citizen?

The state-ownership model, built around national "oil champions," is the model the vast majority of the world is now following. The market economies of the West may not believe in state-owned enterprises, but that doesn't stop countries as different as China and Venezuela, or Russia and Saudi Arabia, from using national oil companies as arms of government in pursuing national interest. Record commodity prices have been a rallying call for resource nationalism around the world, and nowhere has that been more evident than in the oil industry. Rising resource nationalism has led to increasing local resource ownership. Even in free-enterprise America, triple-digit oil prices triggered congressional investigations into price gouging by oil companies and excessive speculation by investors on Wall Street. But it is a lot easier to rein in speculators and hedge funds than it is to control the depletion of the world's oil wells.

When obscene profits are there to be made, most people in the world think it's a better idea that those profits go to national oil champions than to private, and often foreign-owned, companies. That's why the world's largest and most powerful oil companies are increasingly state-owned or state-controlled entities like Sinopec in China or Gazprom in Russia. Not only do they shut privately owned competitors like Exxon out of their own oil patches, but they compete aggressively, often with few market constraints, for access to other countries' oil resources. The big oil companies are rapidly running out of real estate to operate in.

Canada just happens to be one of the few places where the door is still open to them. And it just happens to be next door to the largest oil market in the world, and connected to a pipeline network that can send Canadian oil to American refineries.

The fact is that for a long time oil companies couldn't be bothered to set up shop in Canada. After all, the Canadian oil sands are not a newly discovered resource. The Cree Indians have long used the bitumen to waterproof their canoes. And the process of heating the sands to extract oil was pioneered near Fort McMurray back in the 1920s, with the first limited production flowing about ten years later. It was in the 1960s that the Great Canadian Oil Sands Company, the forerunner of today's Suncor, began commercially upgrading bitumen into crude. The world has known vast unconventional oil reserves for a long time. But who wants a barrel of oil that is worth less than the cost of producing it?

The oil sands didn't make a whole lot of economic sense until as recently as a decade ago, when oil prices started to take off.

Both in Canada and in Venezuela, oil sands development was sponsored by giveaway royalty agreements (less than a third of oil revenue goes into the coffers of the Alberta government; in Norway it is 78 percent). Even with those sweetheart deals, it was hard to attract much interest in such a problematic resource at a time when oil cost $20 per barrel or less. It was just too expensive to get the oil out of the sand.

Deposits of the oil-soaked sand run between 100 and 400 feet thick. Some lie just under the soil and moss and trees (all of which are first scraped away by armies of bulldozers) and are simply shoveled out and hauled to a processing plant, where the sand is heated to intense temperatures to separate the oil-like bitumen from the dirt. But bitumen is far too viscous and dirty to gush—in fact, it is too thick even to flow in any meaningful sense of the word. It must be diluted just to be sent through a pipleline. To get a sense of what bitumen is, think of the black ooze that makes up roofing shingles and asphalt. Bitumen must be processed extensively to become a usable motor fuel. This is done through an extensive cracking process that is much more costly and energy intensive than cracking light sweet crude.

But there is only so much of even this marginal resource near the surface—about 80 percent of the resource is more than 230 feet below the surface, which makes it too deep for open-pit mining. The only way to get at these deeper deposits is a much more complex process of injecting steam as hot as 1,000 degrees Fahrenheit into the sands, which reduces the bitumen's viscosity and allows it to drain away to be collected and pumped to the surface. While less than half of current oil-sands production comes from this process, as much as three-quarters of planned new capacity will involve this far more complex and, again, more energy-intensive extraction, known in the industry as "in-situ production."

Even the more conventional mining operation is extremely energy intensive. An oil-sand operator must typically spend one British thermal unit (Btu) of energy for every three Btu of energy that they extract from the resource. By comparison, for the same energy expenditure you would expect to get 100 Btu of energy from a typical conventional oil well in the Middle East.

But that is the new math as depletion forces us to rely on reserves that can be extracted only with diminishing energy rates of return. With every increase in oil prices, once-marginal sources of oil suddenly become financially viable to exploit— but only at steadily decreasing rates of return, at least insofar as energy is concerned. And in the coming world of scarcity, it is the rate of return on energy investment, rather than the financial rate of return, that will ultimately determine whether energy usage is sustainable in our economy.

TAKE-HOME ENERGY

The concept of an energy rate of return is crucial to determining whether an energy source is economically viable, particularly when it comes to nonconventional oil deposits. The energy rate of return measures the energy you get back from the energy you expend. Obviously, the bigger that return, the better. The one common denominator about all those nonconventional deposits like the Canadian oil sands that the world is now turning to is that you are going to have to burn a lot of energy to get that oil out of the ground. In other words, it has a poor or diminishing energy rate of return.

Thinking about an energy rate of return is like thinking about the difference between the gross and net on your paycheck. What matters on payday is not your gross salary, but what is left after taxes and other deductions have been taken off. You don't get to spend your gross salary—you have to make do with your take-home pay. It is the same with oil, or any other energy resource. There may be 165 billion barrels of oil in the Canadian tar sands, but that is not the same thing as finding 165 billion barrels of light sweet crude in the desert in the Middle East. The "take-home" energy from the oil-sand reserves would be a fraction of what you get from conventional reserves.

There are a lot of deductions when it comes time to calculate the take-home energy of a barrel of synthetic crude from the oil sands: the huge energy costs of a sprawling open-pit mine, the additional energy it takes to refine bitumen as opposed to crude oil (one of the main differences between the two is that bitumen contains less hydrogen than crude and the process in which hydrogen is added is itself very energy intensive, as is the process of producing the hydrogen in the first place). But the biggest energy deduction really stings: if you want to produce a single barrel of synthetic oil from a load of tar sand, you are going to have to burn 1,400 cubic feet of natural gas first. And there is not an inexhaustible supply of cheap natural gas in North America.

So while soaring oil prices can temporarily manufacture attractive financial returns from nonconventional sources like oil sands, they mask the sharp decline in the energy rate of return. Resources that are subject to steadily declining rates of energy return are ultimately unsustainable sources of energy supply.

That's recently become painfully obvious to oil-sand producers as world oil prices crashed as low as $40 per barrel in the midst of a global recession. Recession may have changed the price of oil, but it hasn't changed the cost of extracting it from the oil sands and refining it into a usable fuel source. The same goes for other non-conventional sources like deepwater fields. For a new mining project in the Canadian oil sands, all-in costs can be as high as $90 per barrel. There are much easier ways to lose money than investing billions of dollars in new oil-sand capacity when your cost is going to be well over double the selling price of oil. Without nearly triple-digit prices, the oil is staying where nature put it—in the sand.

Statoil, the large Norwegian oil company, just pulled the plug on its planned $12-billion-dollar development of its Canadian oil-sand properties. Other heavyweights in the region, including Shell, Suncor, Nexen, Petro-Canada and Canadian Natural Resources, have all either canceled or delayed huge projects. Of seven upgraders that were expected to be built to convert the tarlike bitumen into synthetic crude, only the one already well underway by Royal Dutch Shell will go ahead. Why build refineries to upgrade bitumen when oil prices no longer make that economically viable? But even more devastating is the impact of plunging oil prices on production growth.

By the end of 2008, project cancellations had already taken out an expected 1 million barrels of new oil-sand production over the next five years, and the casualty list continues to grow every day that oil prices languish in the $40 range. That drop is nearly 60 percent of the planned production growth in the region. And it makes a mockery of the International Energy Agency's latest forecast of almost 4 million barrels ultimately being pumped out of the region each day (a bigger increase even than the IEA is forecasting for Saudi Arabia).

As it turns out, oil prices aren't the only constraint the resource faces. The other is the price of natural gas. Those 1,400 cubic feet of natural gas it takes to produce a barrel of tar-sands oil aren't much easier to come by than the oil itself and could be put to better use doing something else. If Canadian oil-sand production ever increased from its present level of approximately 1.2 million barrels per day to the 4 million per day that the oil industry forecasts for 2020, Canada would have to cannibalize its natural gas exports to the United States to do it.

That would mean you could fill up your gas tanks but only if you turned the lights off at home. Canadian natural gas has historically supplied about a fifth of American gas consumption. Without it, many US gas-powered generating stations would have to shut down.

Long before that happens, the attendant loss of natural gas supply to the United States may send North American natural gas prices soaring so high that the underlying economics of oil-sand production make even less sense.

Ultimately, natural gas may become so expensive that it becomes more valuable in the marketplace as a fuel source for electricity than to burn as an input into the production of synthetic oil. If that occurs, oil-sand production will become reverse alchemy—turning gold into lead.

What makes using natural gas as an energy input in oil-sand production economically viable right now is the fact that natural gas is a cheaper fuel than oil. The cost of burning natural gas to generate 1 Btu of energy is less than the cost of burning an equivalent amount of oil to produce that same Btu of energy.

It's economically legitimate to burn natural gas to produce synthetic oil from oil sands as long as natural gas trades at a price discount to oil for a common amount of energy in the fuel. But if the price ratio swings around the other way, the natural gas that was going to the oil-sand plants will be bid away by alternative, and financially more attractive, uses for it, like supplying the American market with gas for home heating and electricity generation.

Natural gas prices have remained cheap, at least in North America, relative to the price of oil. That's because oil trades at a uniform global price, while natural gas is bought and sold on regional markets, each with its own price. You can't move natural gas across oceans as easily as you ship oil. Natural gas must first be liquified and compressed at very low temperatures to be transported efficiently by tanker. And while some liquefied natural gas (LNG) is shipped across the oceans of the world, there is simply not enough of it to set a global price.

Natural gas currently trades at an enormous 40 percent discount to oil in the North American market, one of the largest discounts on record. But as we saw only a few years ago when stormy weather knocked out Gulf of Mexico gas production, that ratio can change in a hurry.

For a period of almost half a year, natural gas prices traded at an energy parity premium to oil as the hit to the Gulf of Mexico's gas production was much greater than even its hammering of oil production. Had that price ratio stuck, the huge natural gas requirements would have seriously impaired the economics of oil-sand production.

In theory, nuclear power could provide the energy needed to heat the bitumen and at the same time trim the very substantial greenhouse gas emissions that come from oil-sand production. But given the regulatory and construction lags around building new nuclear power stations in North America, the prospect of nuclear energy in the Canadian oil sands is well over a decade away, leaving the economics of oil-sand production fully exposed to any relative price shift between natural gas and oil prices.

Of course, quite apart from either the price of oil or the price of natural gas, there are some massive environmental issues confronting oil-sand production. The production of a single barrel of oil pollutes 250 gallons of fresh water and emits over 220 pounds of carbon dioxide into the atmosphere. Now multiply that by the over 1 million barrels per day that already comes from the region and you get a sense of the environmental challenge Alberta faces. And now think of almost quadrupling production, as is planned. All of the boreal forest that sits inconveniently atop all that black sand sequestering carbon dioxide out of the atmosphere suddenly becomes "overburden": a waste product that must be removed to allow oil producers access to the oil sands below. In their place they leave an archipelago of tailing ponds—toxic byproducts of oil-sand production and death traps for migrating wildlife.

Whether Alberta would ever sanction the massive desecration of its northern forests implied by the IEA's future production targets is certainly open to debate. While Alberta may not be the greenest jurisdiction in North America, the carbon trail from oil sands has already become a cause célèbre of the international environmental movement.

At the $40- to $50-per-barrel oil prices left by the recession, environmentalists won't have to worry much. No one will be opening up new oil-sand operations at that price or anything close to it. But if the Canadian oil sands—potentially the single largest source of new global supply according to the IEA projections—don't flow, what will? Remember, it's new high-cost supply sources like oil sands and deepwater fields in the storm-ravaged waters of the Gulf of Mexico that we are counting on to replace nearly 4 million barrels per day of production we lose every year from depletion. Without new supply from places like the Canadian oil sands, how are we going to make up that shortfall? And if we can't, how long can oil prices stay below $50 a barrel before the next economic recovery sends them soaring right back into triple-digit territory?

SCRAPING THE BOTTOM OF THE BARREL

When it comes to something as important as oil, we have always gone for the cheapest, easiest, most energetically profitable stuff first. At one time it was the black gold from Texas that powered the economy. When that wasn't enough, it was crude from the giant wells of far-off (but very compliant) Saudi Arabia that kept the economy firing on all cylinders. When Middle Eastern oil became problematic (to say the least), the much more challenging North Sea and Alaskan oil fields were standing by. Now they too are in rapid decline, and we are pinning our hopes on deepwater rigs in the Gulf of Mexico, oil sands in Canada, and other expensive, ambitious projects around the world.

There is a pattern here. We are getting closer to the bottom of the barrel with each move to a new source.

These sources are not really "new." They've all been known about for a long time. What is new is their commercial viability. And it is only the depletion of much cheaper, easier-to-get-to oil that has made such supply commercially viable in today's marketplace. What else is left? If the Albertan oil sands are more enticing than the next option, one can only imagine the difficulties with what lies next on the list of future oil sources to exploit.

Recently, attention has been focused on the Arctic National Wilderness Refuge (ANWR)—a remote 30,000-square-mile federally owned tract of Alaska coastal oil plain on the Arctic Ocean. The reason it was set aside in the first place is that the land is environmentally sensitive and is home to one of the largest caribou migrations in the world. The irony of polluting this pristine landscape to drill for the climate-changing oil that is already wreaking havoc in the Arctic is surely not lost on the people up north whose houses are literally sinking into the melting permafrost.

Environmental issues aside, the ANWR can never be a realistic substitute for the decline in oil production in the Gulf of Mexico or for the chronic delays in Canadian oil-sand production. While Gulf reservoirs are dwindling by the day, we are at least a decade away from seeing a barrel of new arctic oil. Even if the very considerable environmental opposition to drilling in the ANWR is steamrollered and development proceeds as quickly as possible, the US Department of Energy estimates that the first flow of oil from its still unproven reserves could not occur before 2018. And initial first-flow dates in the Arctic are notorious for having to be subsequently pushed back due to construction delays. Even the Department of Energy's best-case view of the region's potential maximum production of just less than 800,000 barrels per day could not be reached until at least a decade after that. Much too little; far too late.

That kind of production is a drop in the bucket when you consider that the United States guzzles over 19 million barrels a day. Arctic drilling is not going to provide even a semblance of "energy independence." And the oil might not even go to the US market anyway. Unlike Russia and Venezuela and other rivals in the world, the United States and Canada don't have national energy companies, so there is really no such thing as American or Canadian oil. The restriction on exporting Alaskan oil was repealed back in the Clinton administration. Whoever produces it gets to sell it to whomever they like. And that means that trampling the ANWR might not even bring down prices at the pumps. Oil is globally priced, so it doesn't matter where it was pumped to the surface—as Canadians know very well. Canada has already achieved the American dream of energy independence, but drivers there pay considerably more at the pump than their US neighbors.

So if the oil resources at America's immediate disposal look more and more like pocket change, what are the prospects for satisfying tomorrow's energy needs? If Canada and Alaska and the Gulf of Mexico can't keep US tanks full, where is the oil going to come from?

The United States is already two-thirds dependent on oil imports, so the quick answer is somewhere else. But where that somewhere else actually is, is far from clear.

Certainly not Mexico, which soon will not be an oil exporter at all. Mexico has already slipped from its historic number two spot to number three on the US supplier list as its huge Cantarell field continues to deplete at an alarming rate. Venezuela, the next petro-power in the hemisphere after Canada and Mexico, sure isn't going to fill the breach. It's unclear whether American motorists have more to fear from the technical problems that seem to always plague the state-owned oil monopoly Petróleos de Venezuela or from the never-ending ideological war between Caracas and Washington. Either way, the outlook is not bullish for filling up your tank in the future.

Venezuela's nationalization and outright expropriation of American oil companies may have been shrewd political moves by President Chávez, and have certainly provided a handsome revenue stream for the government. But the measures have not been without cost. Venezuela's oil production is stagnating, despite the potential of the oil sands in its Orinoco region. The country's exports haven't grown in a decade. And expanding supply to an increasingly vulnerable US market does not rank high on the fiery Venezuelan president's list of energy policy priorities. Falling oil prices may have temporarily humbled Chávez, and sent him back to courting the very multinationals he had recently sent packing. But in an equally telling sign of the times, those jilted oil companies are proving only too happy to hear from him.

But if you are getting qualms about supply from this hemisphere, your anxiety will grow as you cross the Atlantic Ocean and head down the list of countries that send their oil to the States. For all of Chávez's histrionics, Venezuela and its oil industry are a virtual paragon of political stability compared to current conditions in Nigeria, America's fifth-largest supplier. Chávez may not have too many warm and fuzzy feelings about his American oil customers, but at least no one is blowing up Petrólos de Venezuela oil pipelines or rigs every day.

While Nigeria has the capacity to produce roughly two and a half million barrels per day, it never actually does. Anywhere from 500,000 to 1,000,000 barrels per day is typically off line because of attacks on either pipelines or rigs. The simmering civil war even threatens development of once promising offshore fields like Bonga, where drilling platforms have already been repeatedly attacked.

Oil consumers shouldn't expect the Nigerian conflict to end anytime soon. The call for autonomy and sovereignty of the oil-rich region of the country has put the Movement for the Emancipation of the Niger Delta on a head-on collision course with the federal government in the northern part of the country. Since virtually all of Nigeria's oil production comes from the Niger Delta, the secession of that state would leave the country with neither oil nor oil revenue. Which means chances of a negotiated settlement in Nigeria are effectively nil.

Of course there is always Saudi Arabia, home to the world's largest oil production but the source of as little as one and half million barrels to the American market. While Saudi Arabia has recently overtaken Mexico to grab the number two spot on the

US supply list, the only reason for this is that its exports aren't falling as fast as Mexico's. Canada remains the number one energy supplier to its southern neighbor, but as we have seen, the real question with Canadian oil is not how fast its conventional oil production declines but whether its nonconventional oil ever materializes on the scale everyone is counting on.

Whether or not the world's number one producer can actually increase production remains in doubt. In 2008, Saudi King Faisal claimed the desert kingdom needed $75-per-barrel oil to justify any expansion of its production capacity. So the very conditions for Saudi Arabia to supply the world with more oil is reason in itself for you to consume less of it. Without ever-rising oil prices, we will have less and less of it to burn.

CHAPTER 2

DEMAND SHIFT

WHAT'S THE COOLEST THING TO DO IN Dubai on a sunny afternoon? Ski, of course.

Don't let the 100-degree-Fahrenheit temperature outside fool you. There is lots of snow inside, in a complex that is approximately 25 stories high, where the inside temperature is kept at a comfortable 28 to 30 degrees Fahrenheit—just perfect for spring skiing.

Welcome to Ski Dubai, the Middle East's first indoor ski resort. It may be hot enough outside to fry an egg on the hood of the luxury cars gleaming in the parking lot under the desert sun, but every day there is amazing snow for skiing, snowboarding or tobogganing, or, as the ads claim, just plain playing in the snow.

Ski Dubai has five runs that vary in difficulty, height, length and gradient so as to appeal to a wide range of skiers. It even boasts the world's first indoor black (expert) run. The longest run is 400 meters and has a vertical of over 60 meters. And if you are not a skier, you can always just sip a hot chocolate in the crisp, refreshing air.

Okay, maybe it's not Vail or Chamonix, but Ski Dubai's Snow Park gives you 3,000 square meters of year-round perfect snow in the middle of one of the world's hottest deserts.

Being from the land of snow, Canada, I love to ski as much as anyone. If I lived in Dubai I would probably be one of those 3,000 or more people each day who come and look up at the complex's massive roof while riding a chair lift up a run of 100 percent artificial snow.

But since I am a North American driver, and I need to fill my gas tank from time to time, I would greatly prefer that the good people of Dubai come to Colorado or British Columbia to do their skiing and stop burning an obscene amount of scarce oil and natural gas to make indoor artificial snow when the temperature outside hits triple digits.

Ski Dubai and the massive indoor mall where it is located use the energy equivalent of 3,500 barrels of oil a day. Nearly that number of people visit the complex every day. That basically works out to a barrel of oil for each visitor. A standard barrel of oil should power the average vehicle for around 1,000 miles of driving. One day spent on the artificial slopes at Ski Dubai eats up about as much oil as the average American driver burns up in his or her gas tank every month.

I put my two kids, Jack and Margot, on skis when they were three years old, which taught me a long time ago that skiing is an expensive sport. The fuel costs at Ski Dubai, though, make Whistler seem like a bargain. But the skiers and policymakers in Dubai don't care about that cost, and they certainly don't care about what pump prices I pay at my local gas station. Nor should they. After all, it's their oil, and how they want to burn it is their business.

So why should I care how much oil Emiratis use in their leisure time? Because what they do will make my world shrink.

WHY THE LAWS OF ECONOMICS AREN'T WORKING

Of course, we have already seen that the familiar rules of economics don't seem to have been working very well—rising prices have not brought fresh supplies of oil to market. Well, the rules weren't working on the demand side either.

Between 2000 and 2008, world oil prices increased over sevenfold, from $20 per barrel to nearly $150 per barrel. That move surpasses even the enormous price spikes of the two OPEC oil shocks that brought the world economy to its knees in the 1970s. Back then, of course, global demand buckled almost immediately and oil prices came quickly crashing back to the ground. And of course this record increase in oil prices, with the aid of a world financial crisis to boot, has also brought the economy to its knees, but not before oil demand defied the laws of economic gravity for the better part of a decade.

In fact, for most of that decade, world oil demand was growing faster than it did at the turn of the millennium, when oil was only a fraction of the price. That's not the way a demand curve is supposed to behave. Somehow, Adam Smith's invisible hand had been misled or misguided into consuming more oil just when it was expected to burn less.

Why did oil have to skyrocket all the way to nearly $150 per barrel before the world economy stopped growing? Prices half that amount should have stopped global oil demand in its tracks long before the recession hit in late 2008.

For the previous three years, global oil demand had been growing at almost twice the pace of historic norms despite the fact that world oil prices were at record highs. That phemoneon violated yet another sacred truth of economics: the downward-sloping demand curve.

Here's why the laws of economics didn't work. In the world of economic textbooks, there is only one world price for oil. So in that textbook world, higher oil prices lead everybody to consume less oil.

But that's not the world we live in. At least that is not the world as it currently exists for most oil-producing countries in OPEC. Oil prices in those countries don't change. They are politically grounded at a fraction of world oil prices and do not move as world prices climb inexorably higher.

While you may respond to high prices by consuming less energy, your fellow driver on the other side of the world does not have to make such a sacrifice. Economists take it as given that people will make rational choices to make the most of their money. But what is rational behavior in one part of the world can be very different from what constitutes rational behavior in another. Both might make sense measured against their own benchmarks— benchmarks that can be as different as night and day.

If you are a car buyer facing record high gasoline prices, as you did between 2005 and 2008, you probably already made the perfectly rational decision to spend your money on something more fuel-efficient. If you are in the market for a home, chances are you did not look at a huge suburban house that will require you to drive miles and miles to work and back each day. If you drive an SUV, you have probably already quite rationally decided to drive less or to trade your vehicle in for something that does not drain your bank account every time you fill it up. Ridership on public transit had increased noticeably in little more than a year by the time oil prices hit their most recent peak. People make smart decisions in relation to the circumstances they face.

The basic law of demand seemed to be working just fine in North America, Western Europe, Australia and Japan, where consumers were forced to pay the full freight on world oil prices. In fact, soaring oil prices had begun to quash oil demand well before those economies fell into recession. US oil demand had not grown since 2005, when oil prices first crossed over $50 dollars, while oil consumption in Western Europe fell in both 2006 and 2007 despite broad economic growth in the European economy. Price was rationing demand just as economists' theory of the downward-sloping demand curve would predict.

But that hardly explained the world's growing appetite for oil in the face of steady price increases. The answer to that riddle lies with disaggregating world oil consumption regionally to see where oil demand has been coming from and, just as importantly, where it has *not* been coming from. Regional differences aren't just important; they are absolutely critical to understanding what has been driving global oil consumption.

Beneath the surface of seemingly robust growth in world crude demand over the last decade run two very divergent trends. While one part of the world is attempting to decarbonize its economies and wean itself off oil, the rest of the world is burning oil along with other hydrocarbons at a record pace. The global average is simply the netting out of two opposite forces.

Where the demand for oil is the weakest is where historically it's been the strongest. Whereas in the past, economists would have looked at North America and Western Europe to gauge the pace of world demand, now they need to consider the bustling energy demand from developing countries to determine that pace. In many of these countries, it's literally been only a few years since bicycles and rickshaws outnumbered motor vehicles on the road. Yet these are the very countries that will drive future world oil demand.

The OECD accounted for almost three-quarters of world oil consumption in 1990. Today it accounts for barely half. At current growth rates, oil consumption outside the OECD will exceed the group's consumption by around 2012—matching a development that occurred long ago in coal usage.

The very fact that world demand has grown so rapidly this decade means that the opposite forces in global energy markets are far from equal. While oil consumption in the OECD fell in recent years, it fell only at a fraction of the pace that oil consumption in the rest of the world rose.

Whereas world oil demand grew at just under 1 percent over the twenty years following the second OPEC oil shock, between 2000 and 2007 that rate was closer to 2 percent. That's a seemingly perverse result when you consider that oil prices rose by about 600 percent in the same period. But as we shall soon see, prices are not all the same across the different oil markets of the world.

TRAFFIC JAM

If you want to find out what was driving all that demand growth for oil in the developing world, start with the car. Roughly 90 percent of every barrel of oil consumed in the world goes for transport fuel like gasoline and diesel. Look at where car sales are booming, and chances are you will find where oil demand is booming as well.

The emerging economies will account for three-quarters of the new vehicles on the road over the next three decades. Just as oil consumption outside of the OECD will soon surpass consumption within the OECD, the same is about to happen for car ownership. Within five years, the majority of the world's auto fleet won't be driving on the autobahns of Germany or the interstate highways of America but on the crowded roads of India and China, teeming with first-time drivers.

North America may be the land of the car, but that's very much a rearview mirror on where car sales are growing today. Whereas even in the best of economic times, car sales in mature markets like the US and Western Europe grew at a miserly 1 to 2 percent per year, if that, they grew at ten to twenty times that pace in Brazil, Russia, India and China (the so-called BRIC countries, identified as the economic titans of the coming decades by Goldman Sachs in 2001). In early 2009 China's car market surpassed the US market, which is likely to come out of the recession about half the size it went in.

And China was by no means the only place outside of the OECD where car sales were booming. Sales were up 30 percent in Russia and Brazil in 2007, while India also sported double-digit growth rates. Detroit was struggling long before the recession hit, but the world car market had never been stronger. And while the recession has obviously brought those sales gains to an abrupt halt, the long-run prognosis for car ownership in the developing world looks a lot more promising than the outlook for vehicle sales in the developed world. The advent of cars from the likes of Tata in India, which will soon produce a car called the Nano that will sell for as little as $2,500, is making driving more and more affordable all the time.

That in itself is a hugely powerful stimulus to oil consumption because it is the savings necessary to buy a car, rather than the price of gasoline, that often poses the greatest obstacle for households in the developing world to get on the road. But between rapidly rising domestic incomes and even more rapidly decreasing car prices, that obstacle is falling by the wayside. That's why everyone from Nissan to Volkswagen to Ford is joining Tata in India and Chery in China in developing cars with sticker prices under $10,000. The $2,500 Nano is either a miracle or a nightmare depending upon your point of view. And your point of view depends very much on where you live.

Cars in the $2,500 to $10,000 range will allow millions of people in the developing world to drive when they would otherwise never have been able to afford to get on the road. That is no doubt a liberating experience for them, bestowing comforts and convenience that Westerners take for granted. But at the same time it gives each of those new car owners a straw to start sucking at what are already rapidly depleting world oil reserves. And the more they suck through those newly found straws, the less there will be for everyone else—and the higher the price we pay for what we are able to slurp up.

Thinking of driving as a zero-sum game is not something that comes naturally to us. If we think about car ownership in other countries at all, we are probably not fretting that drivers on the other side of the world are pumping what would have been our gasoline into their tanks. But there is a strong sense in which that is exactly what is happening. There can be only so many cars on the road in the world, because there is only so much oil to keep them running. So for every Tata or Chery that rolls out of the dealership on the other side of the world, another car is going to have to be left in the driveway somewhere else—perhaps in *your* driveway.

Where the necessary displacement will occur is in the markets that pay the most for gasoline and where driving behavior is the most sensitive to changes in gasoline prices. Those markets are all found in the affluent OECD countries. And by far the biggest and most price-sensitive of those markets is the American market, where the largest adjustment is already taking place.

Which brings us to the second reason why oil demand has been so strong outside of the OECD, namely, the price of oil itself.

In 2007, OPEC and two nonmember oil-producing countries, Russia and Mexico, collectively consumed 13 million barrels per day of oil. To put that number in perspective, it is almost twice China's 7,500,000 per day and some 1,000,000 barrels a day more than the entire daily oil consumption of all of Western Europe. As a group, it now constitutes the second-largest oil market in the world, second only to the United States.

However, unlike in the American market, where demand is now falling, demand growth in oil-producing countries ranks among the highest anywhere in the world. Oil consumption in OPEC countries rose from 2004 to 2007 at an average annual rate of 5 percent, well over double the world's pace. While the rapid growth in oil consumption among oil-producing countries has largely taken place off most analysts' radar screens, its contribution to world consumption has been almost as great as the much-noted contribution from China.

TOO MUCH OF A GOOD THING

What makes oil-producing countries so thirsty for their own product? It's not about the falling cost of automobiles, as it might be for many first-time car buyers in China and India. Highways in the Middle East sport the fanciest and most expensive cars in the world. No Tatas or Cherys here—more like Bentleys and Mercedes Benzes.

The answer lies not with the cost of buying a car but with the cost of filling its tank. From Caracas, Venezuela, to Tehran, Iran, the key driver behind turbocharged gasoline demand in major oil-producing countries is the price of oil itself. Triple-digit oil prices may have meant $4-per-gallon gasoline prices for American motorists, but you could still fill up for 25 cents per gallon at your local gas station in Hugo Chávez's Venezuela. And motorists were paying only marginally more in Saudi Arabia or Iran, where gasoline sells in the range of 45 to 50 cents per gallon.

If North American or European motorists could fill up at those prices or anything even remotely close, OECD oil consumption would never have fallen, not even during a recession. But motorists in those countries will never see such prices again, while drivers in OPEC countries will always see those prices, no matter how expensive oil is trading on world markets.

OPEC is by no means a homogenous group; it is divided by geographic, cultural, historic and religious fault lines. Yet in countries as disparate as Venezuela, Saudi Arabia and Iran, there has emerged a common domestic political consensus that their citizens should have an unfettered right to consume as much massively subsidized oil as they desire.

In fairness, that attitude isn't found just in OPEC. Where I live, people think similarly about consuming cheap power. I live only 90 miles away from Niagara Falls, the site of one of the world's largest sources of hydroelectric power, so I understand how folks feel about oil prices when they live 90 miles away from Ghawar.

Governments that have tried to raise domestic gasoline prices have quickly felt the people's wrath. Take the summer of 2007 in Iran, for example. The Iranian government attempted to put a collar on runaway fuel-demand growth by raising the price of gasoline. The announcement quickly sparked riots in Tehran as irate motorists took to the streets and even torched a number of gas stations. Fearing that the protests would soon get out of hand, the Ahmadinejad government prudently backed down and rescinded its price hikes.

The notion that Iran, the world's fourth largest exporter of crude oil, should be plagued by domestic fuel shortages may seem incongruous, but fuel shortages are nothing new to Iranians. While Iran may be one of the world's largest crude producers, it is at the same time one of the world's largest importers of refined petroleum products like gasoline. Years of demand growth of 5 percent or more have long surpassed the limited capacity of Iran's aging and increasingly dilapidated refineries.

Every year the Iranian government spends tens of billions of dollars on oil subsidies, at the expense of huge operating losses for the state-owned National Iranian Oil Company. Those subsidies in turn drive explosive fuel-demand growth while at the same time discouraging investment in new refinery capacity. The prospect of spending billions to build refineries doomed to lose money isn't exactly a magnet drawing capital to Iran's oil industry.

American motorists who experienced long lineups at the pumps during the Iranian oil boycott of America may now find it amusing that the Iranian government is worried about supply disruptions or blockades from politically hostile countries due to its growing dependence on gasoline imports. But when it comes to domestic fuel subsidies, no other country comes close to matching the generosity of Hugo Chávez's Venezuela. Massively subsidized oil is not only the foundation for the country's social and economic policy, but plays an increasingly important role in Venezuela's foreign policy. Comrade Chávez subsidizes drivers in politically allied countries like Cuba and Nicaragua and has leveraged Venezuelan oil to forge a quid pro quo with Spain. Even the odd low-income American has been sent a subsidy for home heating oil in the dead of winter.

So great is the popular demand for fuel subsidies that in many OPEC countries higher world oil prices actually *raise* oil consumption, in total defiance of conventional economic logic.

Subsidies turn what otherwise is rational economic behavior on its head.

The higher oil prices climb on world markets, the more the oil-exporting countries rake in. That means more export earnings to chase the 25- to 50-cents-a-gallon gasoline sold at the domestic pumps. And the more oil consumption grows in countries like Venezuela, Iran and Saudi Arabia, the less these countries have left to export to the rest of the world, causing oil prices to rise even further.

If you are Hugo Chávez, the oil market is a virtuous cycle, where self-indulgence leads to self-enrichment. The more you consume of your own oil, the higher the price you get for what's left to export. What you lose from shipping less oil you get back by charging more for what you do ship. But if you are a North American motorist, it's a cycle that will take world oil prices so high that it will eventually take you right off the road—a taxi driver in Caracas is filling up his tank with the gasoline you were planning to use to commute to work.

And it is not just the oil-producing countries that are developing a taste for subsidized oil. Some estimates suggest that as much as a quarter of the world's oil supply is sold to over half the world's population at well below market prices.

The Indias and Chinas of the world can't subsidize their domestic oil consumers to the same extent as Venezuela, Saudi Arabia or Iran, because they can't afford to. They are, after all, oil-importing countries, not oil exporters. But even in countries that rely on the rest of the world for the bulk of their supply, the need to subsidize oil consumers is great.

Energy is the lifeblood of progress. More than anything else, it is energy consumption per capita that separates the millions of aspiring households in India and China from the lifestyle that people in the Western world take for granted. Even at the torrid rate of growth in their energy usage, energy consumption per capita in India and China is only a fraction of what it is in North America. While Americans and Canadians burn up 25 and 29 barrels of oil per capita each year respectively, and Australians and Britons 17 and 10 barrels, the Chinese burn 2 and Indians less than 1.

Those energy consumption aspirations put enormous pressure on governments in developing countries as fuel prices skyrocket out of reach of their growing middle classes. For millions in New Delhi and Shanghai, car ownership is one of the most tangible benefits of the globalization of their economies. And those aspirations resonate loudly in the ears of policymakers there.

That's why both India and China have held their domestic oil prices below the world price. They can't dole out dirt-cheap fuel as the OPEC countries do, but they have shown themselves willing to soften the blow of record oil prices to their citizens and industries by allowing domestic prices to lag behind world markets.

When oil prices recently soared into triple-digit territory, India was spending nearly $10 billion a year subsidizing its domestic oil refineries as they bought oil at world prices and then sold gasoline to Indian drivers at well below those prices. That's a lot of money, but not materially more than the US government spends propping up corn-based ethanol production each year. And developing countries are actually better able to afford this kind of energy spending, since budget deficits in rapidly industrializing countries tend to be much smaller than those in Western economies. For example, the deficit in China is less than 1 percent of the country's GDP, giving their government subsidy room that lobbyists in Washington, where the deficit is running at over 10 percent of GDP, can only dream of.

The very existence of subsidies in India and China and of the far more egregious ones in Venezuela, Saudi Arabia and Iran are all but a guarantee that our recent foray into triple-digit oil prices was no accident. The more that oil is subsidized in OPEC countries, the higher the price of oil that everyone else will have to pay. And that means a smaller world.

BURNING OIL TO KEEP THE LIGHTS ON

As we have already seen, when oil costs $3 a barrel, there are a lot of things you can do with it besides running your car. Ski Dubai, and the 3,500 barrels of oil that keeps the snow on its artificial slopes, may be one of the most grotesque wastes of energy in the Middle East, but unfortunately it is only a hint of what goes up smokestacks in that part of the world.

OPEC burns oil in ways that few other countries could afford to. While the notion of burning crude oil to produce electricity may seem absurd in light of where oil prices have already traded, roughly 30 percent of electricity generation in OPEC comes from oil-fired generating plants.

In some places it is even higher. Whether its through conventional steam, diesel turbine or internal-combustion-driven generating stations, Saudi Arabia gets half of its power from burning oil, while neighboring Kuwait relies on burning oil for about 80 percent of its power. Some OPEC countries—Iraq, for example—get almost all their power from burning oil. Algeria, Bahrain and Qatar burn hardly any oil to generate electricity. Instead their electricity comes from burning natural gas.

Oil-fired or gas-fired, nearly all the electricity in the Middle East comes from burning the very hydrocarbons that the region is expected to export to the rest of the world. There are no coal or hydroelectric resources to speak of, and given the region's volatile history, nuclear is not an option the world is eager to see them pursue.

The world's largest oil-fired power plant, operated by the Saudi Energy Company, produces 3,000 megawatts of power, an output equivalent to three average nuclear power plants in North America. Pending additions will increase that plant's capacity to 5,000 megawatts over the next couple of years as part of an ambitious plan to treble the kingdom's electricity production by 2020. All of the planned increases in Saudi power capacity will come from burning oil or natural gas. And none of that fuel will be burnt at anything remotely close to world market prices.

Instead, that Saudi utility company is getting an even bigger price break than Saudi motorists get at the pump. Since the spring of 2006, a royal decree has fixed the price of oil for Saudi power plants at the set price of .46 cents per million Btu. Since a barrel of oil contains 5.8 million Btu, the royal decree translates into a fuel cost under $3 per barrel for the Saudi oil-fired utility. That in turn works out to about 7 cents a gallon.

Power production in the Middle East already uses up about a third of a million barrels of oil per day. That amount will grow rapidly, due to economic, demographic and fuel-mix considerations.

Population growth rates in the Middle East, often as high as 2 percent per year, are among the highest in the world. Since 1950, the population has quadrupled to 432 million and is projected to grow by half again over the next forty years. Not only is population growth rapid, but per capita energy consumption is boosted by the billions of petrodollars that triple-digit oil prices have brought to the red-hot economies of the region.

Power demand in Dubai, home of our favorite Middle Eastern ski hill, is growing at a rate of 15 percent per year. While the United Arab Emirates has the fifth-largest natural gas reserves in the world, it faces domestic natural gas shortages due to long-term export contracts. As a result, the Emirates are switching from burning natural gas to burning oil for electric power generation. Just wait to see what it will have to burn when the UAE hosts its first World Cup ski race.

All of a sudden massive oil subsidies in the Middle East become massive power subsidies. And just as underpriced gasoline is over-consumed, so underpriced electricity is also overconsumed. Saudi households pay about 1 cent per kilowatt-hour, roughly one-tenth of what a typical American or Canadian household would pay and about one-twentieth of what you would pay in the UK. And that's pretty well par for the course for electricity rates among most of Saudi's OPEC neighbors. Not surprisingly, power demand across the Middle East is expanding at between three and four times the pace of power demand growth rates in North America or the rest of the OECD.

Surging power demand growth rates have made the Middle East the highest per capita users of electricity in the world. Per capita power use in Kuwait now surpasses even the American standard and more than doubles consumption in the UK. In Dubai, power consumption is already twice the per capita power consumption in America.

But blackouts and brownouts are becoming increasingly frequent as generating capacity remains hard pressed to keep abreast of runaway demand growth. Rolling power interruptions have become commoplace in Kuwait, particularly during recent summers when air conditioners ran around the clock to beat the heat. Blackouts are frequent in parts of Saudi Arabia, the United

Arab Emirates and Oman. Even Abu Dhabi has been forced to redirect some of its natural gas from oil-field injection to power plants in order to avoid blackouts. The diversion comes at the cost of lower oil production, since natural gas is used to maintain well pressure and hence oil flow in depleting fields.

Power blackouts rarely go down well with the general populace, particularly in places like the United Arab Emirates where average per capita incomes can run in the hundreds of thousands of dollars. What good are all those petrodollars if the lights don't work? When people have money, they expect to be able to consume all the power they want. And most things people want in a scorching desert require a lot of power, even if they aren't into skiing.

Middle East governments might cut back on power subsidies (something they have never been able to do), but they may still find it difficult to restrain energy demand growth. There is yet another powerful dynamic driving the Middle East to burn more and more of its own oil.

TURNING OIL INTO WATER

Saudi Arabia isn't just running out of oil—it's running out of water. So is the whole region. And the water depletion problem is far more pressing and urgent to the people on the ground than any oil depletion problem that they may face down the road.

Their inheritance of fresh water, like their oil, was laid down eons ago in underground aquifers. In countries like Saudi Arabia, fresh water in underground aquifers is already down 50 percent from the levels of the mid-1990s. Yet the Saudi population and economy are both growing at an explosive rate, and so, naturally, is the demand for water.

Current water use in Saudi Arabia is seven times sustainable levels. And some of Saudia Arabia's neighbors are running through their water supplies at an even more alarming rate. In Dubai and the rest of the Emirates, water use is running fifteen times higher than natural replenishment, while in Kuwait it is running at over twenty times sustainable levels.

Take into account the fact that nearly two-thirds of all the water in the Middle East is used for agriculture, and it becomes clear that many Middle Eastern nations are facing the difficult choice between water and food.

Relentless population growth puts more and more pressure on countries like Saudi Arabia to make the desert bloom. But making the desert bloom requires water, and the same population expansion that pushes for greater domestic food production in the Middle East also pushes for more drinking water.

Saudi Arabia drained much of its aquifers for irrigation in the 1980s and 1990s so as to become self-sufficient in wheat. But faced with growing water shortages that can only worsen with rapid population growth, the country is now cutting back on irrigation.

Of course, by cutting water supply to agriculture, the country will grow less food, trading food self-sufficiency for water. To that end the kingdom is seeking investment in other countries' agricultural sectors to secure future food supply for its own rapidly expanding population. It has already begun active investment in rice farms in Thailand and is looking at places like Sudan and Pakistan for fertile land where it can grow crops and export them back home for domestic consumption.

Saudi Arabia doesn't have a Great Lakes or a Mississippi River that it can rely on to meet its water demands. But what Saudi does have is lots of oil and natural gas.

And there is no shortage of saltwater in the Gulf of Arabia. Together, they provide a way out of the country's looming water crisis: desalination. And it is a way out that other countries in the region will likely follow as well. But desalination is a solution whose energy requirements will make the energy costs of Dubai's ski hill look trivial by comparison.

Desalination is an inherently energy-intensive process. One technique, reverse osmosis, forces saltwater through a semipermeable membrane under enormous pressure to remove impurities. Flash distillation, an alternative process popular in the Middle East, vaporizes seawater by passing it through a drum of reduced atmospheric pressure then reliquifying it once it is free of impurities. Either process carries huge energy requirements.

Saudi already accounts for 60 percent of the region's desalination plants, including the world's largest facility, Shuaibah 3. When it reaches full capacity in 2009, it will produce 31 million cubic feet of water per day. Around the same time, Saudi will open the Ras Al-Zour and Al Juball plants, with capacities of 28 million and 26 million cubic feet per day respectively. Together, these plants will boost the country's desalination capacity by 80 percent.

Shuaibah 3 will be powered by one of the biggest generation plants in the world, sending a steady stream of oil that costs less than $3 a barrel up its smokestack. Every year Saudi Arabia depletes about 700 billion cubic feet of its natural water in aquifers. In order to desalinate a compensating amount of salt water, it will have to burn roughly 300,000 barrels of oil or the equivalent in natural gas.

The World Bank estimates that the Middle East will need roughly another 50 to 60 billion cubic feet of water annually over the next ten to fifteen years to meet the region's burgeoning water demand. Desalinating that immense volume of water could ultimately require the use of a million barrels of oil per day.

That's a million barrels a day that won't be fueling the world's cars and trucks. To put that in perspective, the largest oil field discovered in the North Sea in the past twenty-five years, Buzzard, has a daily output of less than a fifth of that. All of the Alaskan oil fields together pump out considerably less than a million barrels a day. Peak water will hasten peak oil.

THERE ARE BETTER THINGS TO DO WITH OIL THAN BURN IT

Desalination plants are not the only new competitors jumping to the front of the line looking for their share of the Middle East's increasingly scarce oil and natural gas.

That part of the world is rapidly becoming the center of the global petrochemical business, with its huge requirements for fossil fuels. And that's all about fuel costs. Why produce ethylene from $5-per-mBtu natural gas in North America when you can produce it from less-than-$1 gas in the Middle East? The allure of cheap feedstocks has attracted over half of the world petrochemical industry's investment to the region in recent years.

Plastic is made from propylene, which is a basic petrochemical that can be created by cracking either natural gas or oil. There is a whole lot more economic value added in producing products like plastic from oil than there is from turning oil into gasoline or diesel. This means that the Middle East's rapidly expanding petrochemical industry will soon get first claim on any oil supply from the region, ahead of the needs of oil refineries elsewhere around the world.

That's not politics, like in 1973 or 1979—just plain economics.

You can make five times as much by turning oil or natural gas into a petrochemical as you can from selling it as a transport fuel. And you create a whole lot more manufacturing jobs in plastics and chemicals than you do in oil refining.

Ultimately, the Middle East may become the center of the world's chemical industry. Abu Dhabi already has a thriving plastics industry and is positioning itself to become a major supplier to the rapidly expanding Chinese auto industry. Borouge, the Abu Dhabi–based plastics manufacturer, has just announced plans to build the world's largest integrated chemicals and plastics plant at Taweelah, near the border between Abu Dhabi and Dubai.

Saudi Arabia has no less ambitious plans for its burgeoning plastics and chemical industries. While Saudi Aramco has focused on refining and on oil-based petrochemicals, Saudi Basic Industries Corporation (Sabic) has gone after higher-value-added fine chemicals that are made from natural gas feedstocks. Through aggressive acquisitions of Western companies and technology as well as through joint ventures, Sabic is rapidly becoming a player in the global chemical industry and intends to become the dominant global player within a decade. It has an excellent chance of succeeding.

In fact, long after triple-digit oil prices render the internal combustion engine and gasoline obsolete, oil will still have value as a petrochemical feedstock. In the end, that may be the only lasting hydrocarbon legacy in the region.

THE VIEW FROM THE PEAK

What makes the growing domestic thirst for oil in OPEC countries so ominous to Western consumers is that it is coming precisely at a time when oil production in those very countries is either peaking or declining. It is one thing for demand to grow rapidly in oil-producing countries when production is also expanding. But it is a whole other ball game when explosive internal demand for oil occurs when production is stagnant, or even worse, in decline.

The tandem of soaring domestic consumption and stagnant production is a recipe for world shortages since it means that major oil-producing countries may soon be cannibalizing much of their own export capacity, ostensibly at the expense of world oil supply. Even more disconcerting is the prospect that this displacement of exports is likely to occur in the very countries that Western oil consumers have been repeatedly assured will provide most of the world's future supply growth.

The cannibalization of exports is already apparent. Despite the still overriding importance the world attaches to OPEC production, exports from the cartel hardly grew at all over the first half of this decade and will likely fall over the second half. Had it not been for the simultaneous surge in Russian oil exports, the world would have seen triple-digit oil prices a lot sooner than it did.

And now exports have begun to decline. Oil exports from the Middle East actually dropped by almost 700,000 barrels a day in 2007, led by a 600,000-barrel-per-day drop from the region's largest exporter, Saudi Arabia. While the Kingdom has claimed that the decline in its and the region's exports are temporary, that dropoff follows on the heels of stagnant exports over the previous five years and of growing struggles to simply maintain current production levels. Yet the official party line from Saudi is that it will boost its production and exports significantly over the next decade.

The reality is likely to be very different. While production may not fall in the Middle East, exports are likely to drop by nearly a million barrels per day over the next four years due to increased diversion of oil to meet soaring domestic demand. And even bigger cuts are likely down the road, once depletion in many of the world's oldest producing oil fields accelerates.

The Middle East has been the workhorse of world petroleum production for almost a century, and the cumulative draw from the region is beginning to show. Huge fields like Burgan, in Kuwait, are now a shadow of their former selves, with production rates a fraction of what they were at peak. And there is growing speculation that Ghawar, the world's largest field and still the home of over 50 percent of Saudi Arabia's production, may soon start to produce less as well. More and more of what comes out of the ground at Ghawar is diluted with seawater because engineers are pumping the reservoirs with seawater to increase the pressure that forces oil to the surface. There is no reason to try to revive an oil field that isn't flagging, so the implication is clear: the biggest oil field in the world's largest oil producer is in decline.

At the same time, production in other Middle East oil heavyweights like Iran is a far cry from what it was at peak. Iran's current production of just less than 4 million barrels per day is a third lower than during the time of the last shah; Iran's exports to the world have fallen by about 2 million barrels per day since the Iranian Revolution in 1979.

The story isn't much different in the rest of OPEC, outside of the Middle East. Production in Venezuela, one of the organization's founding members, has fallen steadily, starting even before Hugo Chávez effectively nationalized the country's oil industry. Conventional oil production in the country has fallen there by as much as a million barrels per day over the last decade. Ditto for Indonesia, where crude production is now only half of the level it was thirty years ago.

Indonesia is a particularly intriguing example because it represents Western consumers' worst fears. While up until recently technically an OPEC member and an oil-producing country, Indonesia has been a net oil-importer since 2004. Its million barrels per day of production is now almost 20 percent lower than its daily oil consumption, hence it must now draw upon supplies from the rest of the world when a decade ago it was a net exporter. Recently, the Indonesian government withdrew its forty-six-year membership in OPEC.

Until very recently, the world was betting that if OPEC dropped the ball, Russia would be there to pick it up, just as it has done for the better part of the last decade. Russia is not only the world's second largest producer of oil, but unlike Saudi Arabia, its own domestic oil consumption took a downturn as the economy imploded in the wake of the collapse of the Soviet Union. Russian oil production soared from 6 million barrels per day to over 9 million, and almost all of that production increase was available for export to world oil markets. That was a double win for European energy consumers dependent on Russian oil.

While the world has waited with bated breath following every OPEC supply announcement, it's been growing exports from Russia, not from Saudi Arabia, that have allowed motorists around the world to fill up their tanks. Since 2000, Russian oil exports are up 60 percent, effectively filling the breech created by OPEC's cannibalizing of its own production. In the process, Russian, not OPEC, exports accounted for almost half of the entire increase in world supply from 2000 to 2006.

But now Russia's exports are succumbing to the same forces that have curtailed OPEC's. Recent announcements from the Russian oil industry have acknowledged that production did not grow in 2008 and would likely not grow at all over the near term due to depletion in the mature oil fields of the western Siberia basin. At the same time, efforts to expand production in much less developed eastern Siberia have been mired in controversy and mammoth cost overruns.

But as is the case elsewhere, depletion isn't the only factor constraining exports. When oil prices are high, the economy of the world's second-largest producer, Russia, starts to heat up. Between 2004 and 2007, the Russian economy grew at a not too shabby 6 percent annual rate and boasted annual vehicle sale increases in the 20- to 30-percent range. Guess what happened to Russia's once anemic demand for oil?

As is already the case with OPEC, Russian oil exports will soon start to decline, possibly by as much as a half a million barrels per day over the next half decade. Instead of offsetting the decline in exports from OPEC as Russia has done over the first half of this decade, the northern energy giant will now exacerbate the loss of OPEC exports with curtailments in its own oil shipments to the rest of the world. And Russia's pursuit of greater coordination with OPEC will do nothing to bring prices down.

Yesterday's solution just became part of tomorrow's problem.

Together, the two largest oil producers in the world, OPEC and Russia, will not only be unable to meet any future demand growth but between them will be cutting back nearly one and a half million barrels per day of supply to world oil markets by 2012, if not sooner. But the coming supply crunch doesn't end there. Another oil-exporting country is likely to lose even more of its exports than either Russia or OPEC over the next five years.

Exports from Mexico, until very recently the number two supplier behind Canada to the huge US oil market, will collapse over the next couple of years. They already fell 20 percent in 2007, but the meltdown has just begun. While Mexico is unlikely to become a net oil importer like Indonesia has, it will nevertheless cease to be an oil exporter of any consequence. Its current one and a quarter million barrels per day of oil exports, almost all to the American market, will shrink to something more like a quarter million barrels per day or less.

Mexico's problem is its major oil asset, the vast Cantarell field under the Bay of Campeche in the Gulf of Mexico. Currently the third-largest producing field in the world, it is home to about 40 percent of Mexico's total oil production. We have already seen that deepwater oil fields have particularly steep decline rates, and Cantarell seems to be living out a worst-case scenario. It has already lost over a million barrels per day in production and is expected to lose at least half of its remaining million barrels per day over the next five years. Juxtapose that rapidly depleting supply outlook against the growth in Mexico's own gasoline consumption, and you can see that pretty soon all of the country's depleted oil production will go to filling the tanks of Mexico's ever-increasing fleet of cars.

Russia, OPEC and Mexico together account for about 60 percent of world oil production, with combined output of just over 44 million barrels per day and total exports of roughly 35 million barrels per day. While the group as a whole should be able to maintain total production rates close to current levels, growth in domestic consumption will cut exports by almost 3 million barrels per day.

Unless the recession that began in 2008 turns into a decade-long depression, the loss of 3 million barrels per day will send oil prices soaring again. Price increases must destroy enough oil demand in North America and Europe to allow consumption to continue to grow in oil-producing countries like Saudi Arabia and Venezuela and in places like China and India as well.

It may seem unfair that Emiratis are carving turns on fresh snow in a huge refrigerated bubble under the scorching desert sun while we are looking for ways to cut back. Or that car dealerships in the developing world were only recently doing a booming business while Detroit's Big Three (GM, Ford and Chrysler) were shuttering plants. But those are the fundamental shifts in demand and supply that are reconfiguring our world. And when the dust settles from the movements of those tectonic forces, the shift in energy supply and energy demand is bound to make our world a lot smaller.

CHAPTER 3

HEAD FAKES

THE PINTO, THE GREMLIN, THE CHEVETTE, THE K-CAR—Detroit's engineers and designers have a lot to answer for. When OPEC shut off the spigots in the 1970s, the people who brought the world the American dream in the form of hulking, throbbing rocketships of chrome and steel started filling their showrooms with cars that we look back on with a mixture of amusement and embarrassment. Like a lot of things we remember wincingly, though, these cars seemed like a good idea at the time.

At least from an economist's perspective, if not from a car lover's, they *were* the right idea. When fuel prices soar, you want more fuel-efficient cars.

I remember trading in my dad's old hand-me-down Buick LeSabre for a Toyota Corolla back in the early 1970s—times that in many ways resembled today's environment. It was quite the transition.

I sailed into the dealership under the power of a muscular 5-liter V8 engine, and drove away revving a little 1.6-liter in-line four. Somewhere in Japan, I thought, a lawnmower was missing its engine. And I had coughed up an extra $500 for the more "powerful" engine option—the standard model came with a 1.2-liter motor. (I saved some money on the transmission, though, in the hope that a four-speed standard gearbox would help me wring a little more power out of the tiny engine. Unfortunately, I had never driven a standard, and I burned out a couple of clutches before I had quite mastered the technique. But I haven't looked back! I've never driven an automatic transmission since.) The car cost only about $2,500 back then, more or less the same price as Tata's Nano today.

Of course most of us really didn't want to drive a tiny Japanese car. We would have rather kept our bigger, more powerful and more comfortable vehicles, if we could have afforded to fill them up. But the Middle East had just turned off the tap and the price of gasoline was spiking; pundits were talking about the end of oil; and the North American economy had just slipped into one of the deepest recessions since the Second World War. So just like drivers in Venezuela respond to low pump prices by burning more fuel, we took a look at the cost of driving our Motor City V-8s and made our way over to the brand-new Datsun and Toyota dealerships that were opening up across North America. (Better that than a Pinto.) In the UK, the Mini Metro showed that British engineers could build a popular car around a 1-liter engine.

We couldn't control supply—clearly, OPEC was in control of that back then. So we tried to control demand. We may have missed the raw power and smooth ride of American steel, but we didn't miss the beating we would have taken at the pumps.

And it wasn't just people shopping for cars who did what they could to control demand by looking for fuel efficiency. Governments started to mandate it. In the United States, where the transport sector accounts for about 30 percent of total energy usage in the economy and about 70 percent of the usage of oil, the government brought in the so-called CAFE (corporate average fuel economy) standard in 1975, forcing carmakers to more than double their fuel efficiency, from about 13 miles per gallon to 27 within ten years. Many governments lowered highway driving speed limits in an effort to legislate better fuel economy on their roads. President Carter went so far as to say that kicking Americans' addiction to oil was "the moral equivalent of war."

The effects of high prices were felt around the world—from state governments in the US asking citizens to turn off Christmas lights to the Danish government imposing jail terms for people who exceeded their electricity rations, and from the UK government asking Britons to heat only one room in winter to a ban on flying, boating and driving on Sundays in many European countries.

Grants were given to homeowners to install new and better insulation in their homes so as to cut down on their consumption of home heating oil during the winter. In North America there was a massive expansion of natural gas pipelines to encourage households to switch from high-cost home heating oil to natural gas, which was abundant and cheap back in those hazy days.

A lot has changed since then. But then things started looking pretty familiar when we got a glimpse of what a world of triple-digit oil prices looks like. In North America, carpooling is making a huge comeback, scooter sales are soaring as rapidly as SUV sales are plunging, and an increasing number of lanes on our highways are now being reserved for vehicles carrying multiple passengers. Moreover, the dawning public awareness that dependence on scarce, expensive foreign oil and dependence on cheap, abundant foreign oil are two very different things has led to some ambitious and urgent plans to switch to alternatives like ethanol. In other words, if we can't control the supply of oil, we had better roll up our sleeves and get to work on controlling the supply of something nearly as good as oil.

Behind all these initiatives are the implicit beliefs that by becoming more efficient in our use of energy we will use less of it and that if we just get serious and throw some money at technology we will discover an alternative fuel source. But is either a realistic premise? Does greater energy efficiency lead to actual reductions in energy usage? And are we likely to be able to grow the fuel that we used to just go out and find?

These are no longer academic questions. When conventional oil is becoming scarce and nonconventional oil is the energy equivalent of pocket change under the sofa pillows, questions don't get much more urgent.

Anyone who remembers the Pinto knows that even really smart people sometimes respond to complex problems with solutions that look pretty stupid in hindsight. Are we implementing exactly that kind of solution right now?

THE REBOUND EFFECT

Common sense is not without its virtues. All those people who bought fuel-efficient cars in the 1970s did save money, and all that public and private investment in new energy-saving measures really did allow us to become more fuel efficient. Since 1975, energy consumption per dollar of GDP in the US economy has dropped by nearly 50 percent. We are far more efficient than we were when I traded in the old LeSabre.

But don't be fooled into thinking that means we use less energy. Because in fact we use more than we ever have in the past. Much more.

Energy consumption is up over 40 percent despite cutting the energy input per unit of GDP in half. That's why energy intensity targets, commonplace in most countries' energy strategies, have patently failed to restrain energy consumption and the carbon trail that follows from it.

Following the OPEC oil shocks, a few renegade economists, like Daniel Khazzoom in North America and Leonard Brookes in the UK, argued that improvements in energy efficiency would lead to an unexpected and unwelcome result—increased energy consumption. Their warnings seemed to fly in the face of all those government policies to encourage greater energy efficiency. Needless to say, their analysis was not well received by either policymakers or environmentalists at the time. Khazzoom's life was actually threatened when he testified at an environmental hearing that the Province of Quebec held before proceeding with its massive James Bay power project.

The "rebound effect," as it has come to be known, poses a disturbing assault on both intuition and conventional wisdom. Yet its paradoxical conclusion is based on economists' standard theory of demand.

As improvements in energy efficiency lower the price of consuming energy, more energy will be consumed, as predicted by the economists' standard theory of a downward-sloping demand curve. That is, while efficiency reduces demand for energy, reduced demand in turn reduces the price of energy. The effect is that you end up getting more energy for the same price. So you naturally end up using more.

And using more energy means more economic activity—more driving and building and manufacturing and, inevitably, shopping. So if an efficient economy can consume more energy for the same price, it also gets more economic growth for the same price—and that means still more stimulus to energy demand from a stronger, growing economy. In other words, cheap energy makes the economy grow, and a growing economy is greedy for more energy. When economic growth outstrips the rate of improvement in efficiency, the result is a very powerful rebound effect.

The concept was described over a century ago by the British economist William Stanley Jevons. Jevons observed that after the huge efficiency gains following the advent of James Watt's steam engine, coal consumption initially dropped, then rose tenfold between 1830 and 1860.

The same phenomenon occurred with efficiencies in steel production in that era. The Bessemer process for producing steel was one of the greatest fuel-saving innovations in the history of metallurgy, but its ultimate effect was to increase, not reduce, the industry's demand for fuel due to the subsequent surge in steel production.

While each ton of Bessemer steel or increase in horsepower of James Watt's steam engine might require less fuel than before, skyrocketing increases in the demand for steel and power overwhelmed the efficiency gains, leading to significantly greater fuel consumption.

And that is exactly what has happened to your car.

When high prices sent the automotive engineers to their drawing boards to come up with more efficient vehicles, the results were nearly immediate. From lighter materials to fuel injection and turbocharging and multiple-valve cylinders, the automakers responded quickly to the OPEC crisis by coming out with technologies that would give drivers more power for less fuel. These new engines improved mileage considerably. Since 1980, the average mileage per gallon of gasoline in the United States has improved by an impressive 30 percent. Technology compensated for the rising cost of fuel, just as economic textbooks would have predicted. In fact, the transport sector has become the economists' poster-boy example of how prices influence technological change.

But roll the clock forward to a few years after the second oil shock. The advent of new fuel-saving technology didn't follow through where it mattered the most, which is ultimately not how much fuel it takes to drive a mile but how much fuel your car actually uses over the course of a year. On that score, absolutely nothing has changed. Your average vehicle on the road in America consumes just as much gasoline today as it did three decades ago when its engine was 30 percent less efficient than the engine in your car today.

What all the new automotive technology allowed carmakers to do was to squeeze more power out of the same amount of fuel. That meant that all things being equal, a driver could move his or her car down the road at the same speed and use less fuel to do it. But all things are not equal. Instead of propelling that car down the road at the same speed, carmakers realized that the new technology would allow them to move a bigger car for the same amount of fuel, and they could make it move faster, too. The temptation to turn that extra power into more speed and size was irresistible. Instead of using the lighter new materials to develop less energy-intensive vehicles, the car companies often chose instead to use these advances to simply put bigger, faster models in their showrooms.

The efficiency paradox has allowed that four-cylinder Corolla to somehow morph into a huge honking sport utility vehicle. The number of light trucks (a category that includes SUVs, vans and pickup trucks) rose by 45 percent between 1995 and 2005, seven times faster than passenger cars. In a blatant exercise in energy obesity, light trucks account for 80 percent of the growth of the number of vehicles on the road in the US since 1985, becoming without doubt the vehicle of choice for the average American family.

My dad's Buick LeSabre was no Toyota Corolla, but it was modest compared to the legions of Yukon Denalis and other over-sized SUVs cruising out there today. All the technology under the hood gives them better fuel economy than similarly sized trucks from the 1970s, but all the advances of the last four decades don't change the fact that when oil spiked into triple-digit territory, it cost well over $100 to fill one of those monsters' tanks. There is the rebound effect in spades. Your engine is more efficient, but you are burning more fuel.

And even if you aren't driving an SUV, chances are you are driving a vehicle with a whole host of energy-consuming features that were once costly options. Power windows, power sliding sunroof, power side mirrors and, most of all, air-conditioning are now pretty well standard on most vehicles in the developed world. And every day vehicles come out with new power-sucking features, such as onboard computers and entertainment systems. All of these energy-using features are just further examples of how the falling cost of consuming energy has led us to consume ever more of it.

Add it all up, and the vehicle idling beside you on a North American street is probably less efficient than a 1908 Ford Model T.

So much for the great benefits that energy-saving technology has bestowed. From a conservation point of view, the bad news doesn't end there. America's gasoline consumption is not just about average fuel mileage per vehicle. It's also about how many vehicles are on the road. Here, too, we hear the loud echo of the rebound effect.

Improvements in fuel efficiency have allowed more people to drive by lowering the cost of operating a car. Today, there are some 130 million more drivers on the road in the United States than there were in 1970. Over the past decade, the number of vehicles on American roads has grown twice as fast as the pace of household formation. Whereas in the 1970s most American families expected to own a single motor vehicle, today most households have a second car and some even a third.

And not only are there more cars on the roads, we are driving them more. In 1970, the average American car was driven only 9,500 miles per year. By the time of the new millennium, it was driven over 12,000 miles.

More cars, bigger cars, driven more. That's what all the improvements in fuel technology have got us. The result is that we are as gasoline dependent today as we were in the midst of the past two oil shocks.

We could have been developing increasingly efficient vehicles all this time. But when oil prices came crashing down in the 1980s, so did the enthusiasm for efficiency. By 1985, the date by which the automakers were meant to reach the 27-miles-per-gallon target mandated ten years earlier by the CAFE legislation, only Chrysler had hit the mark. The others complained that the CAFE standards were too stringent and expensive to comply with. They managed to get the target rolled back to 26 miles per gallon—not a huge slip, but one that proved hard to recover from. In 1990, the US Senate rejected a bill that would have updated the CAFE standard, raising it by 40 percent. In 2003, the government again failed to tighten the standards. And in any case, the mighty SUV, so essential to the bottom lines in Detroit, was exempt from CAFE.

The result has been a huge rebound not only in miles driven but in the size and power of the vehicles that we drive. Decades after the work of increasing efficiency began in earnest, the American vehicle fleet still consumes about 12 million barrels of oil a day.

FREQUENT FLIERS

What happened on the roads also happened in the skies.

Aircraft engineers managed to squeeze even more efficiency out of design improvements than their peers in the automobile industry when the price of jet fuel took off in the 1970s. Again, this is what an economist would predict, particularly since the cost of air travel is so fuel-price sensitive. Technology responding to price signals: textbook economics.

Fuel consumption per mile flown has improved by more than 40 percent since 1975. Not only did aircraft manufacturers change engine technology in response to higher fuel prices, but they also modified aircraft design to make it more fuel efficient. In direct response to the OPEC oil shocks, airplane manufacturers started to widen their planes in an effort to reduce the number of flights necessary and hence to cut down on soaring fuel costs. Efficiency measures were out in full force in the airline business. But as we saw with autos, the best-laid plans of corporate planners and engineers were once again stymied by the powerful rebound effect.

What wasn't foreseen was that wider aircraft lowered costs per passenger and in turn led to lower airline prices, which consequently induced an increase in air travel by the public. Instead of lowering fuel consumption by reducing the number of flights, wider aircraft—through lower operating costs and ticket prices—led to an increase in the number of flights. The result, of course, was an increase in the consumption of aviation fuel.

As in the case of motor vehicles, increases in energy usage quickly outstripped gains in energy efficiency by a ratio of four to one. As technology allowed the cost of flying to fall despite higher fuel costs, more people started to fly. More travelers means more planes in the air; more planes means more jet fuel burned.

Overall fuel consumption in aviation has risen by 150 percent in the United States.

The engineers did their jobs. And their innovations accomplished what they were meant to—namely, allowing us to use energy more efficiently. But in neither case did that efficiency lead to the conservation of any energy. Like James Watt's steam engine or Bressemer's energy-saving steel process for coal use, improvements in the energy efficiency of vehicles and airlines have simply meant more people on the roads and more people in the skies.

THE REBOUND EFFECT AT HOME

The same perverse patterns between improved fuel efficiency and increased fuel usage found throughout the transportation sector can also be found in the average home, where roughly another 20 percent of energy usage in the economy occurs. Improvements in thermal insulation and in the energy efficiency of major household appliances, especially energy-sucking furnaces and air-conditioners, have all helped to make major gains in energy efficiency in the home.

Today, almost every major household appliance in the developed world must meet some minimum energy-efficiency standard. None of those standards existed in the 1970s. Moreover, heating and cooling systems are as much as 30 percent more efficient in their energy usage than comparable systems three decades ago. But have all these efficiency gains actually reduced energy usage in your average American house?

Not even close.

In fact, the opposite is true. As in the case of vehicles, usage has grown much faster than efficiency. Whereas the average air-conditioner is 17 percent more efficient than it was in 1990, the number of air-conditioners in American homes is up 36 percent, as what once was considered a luxury item has become a standard feature across North America. Putting in an efficient air-conditioner may mean you use less electricity than you would if you had gone with an energy-sucker, but even the most modern unit is still going to use more power than if you had none at all.

Most importantly, look at what's happened to the average size of a North American home. Since 1950, the average American home has grown from 1,000 square feet to almost 2,500 square feet today. That's two and a half times bigger. We are certainly not two and a half times more efficient. Again we can hear the echo of the rebound effect. The average Australian home is about the same size, and homes in Canada and New Zealand are just slightly smaller. (The average home in the UK or Germany is about one-third as big.)

Hence, while efficiency improvements have allowed us to heat, cool and light space much more cheaply than ever before, we now have much more space to heat, cool and light. By super-sizing our homes we have consumed all those energy-efficiency gains in the form of ever bigger heating and power requirements. Moreover, we have filled these bigger houses with things that need to be plugged in. Telephones, for example, used to require only the low-voltage current that came out of the phone jack. Now they need to be plugged into a receptacle, along with all the other things that have those little black transformers that proliferate in power bars across the continent. Our cellphone chargers and halogen lighting and a million other little things keep the electricity meter turning even when they are not performing any task. And those are just the small things.

The big things add up even faster. Plasma televisions use as much as four times as much electricity as the old-fashioned models they usually replace. When people in the UK started rushing out to buy new flat-screen TVs in anticipation of the last World Cup of soccer, authorities quickly realized that these sports fans could pose a real problem for the grid: turning on all those plasma screens at the same time would draw 2.5 gigawatts, or the equivalent of the output of two nuclear power stations. Though Britain did not go dark, when a soccer game strains the grid to capacity it's pretty clear that our TVs are not a trivial drain on our power supply. Add in all the Blu-ray players, Xboxes, wireless routers, alarm systems and so on, and it is not difficult to see why the rebound effect is felt so strongly at home.

In Canada, some people actually install electric driveway heaters to relieve them of the chore of shoveling snow in the winter—clearly there is a lot of low-hanging fruit when it comes to scaling back our energy use. But we are deceiving ourselves if we think that tweaking our lifestyles a little will make them sustainable.

HOW TO BE EFFICIENT WITHOUT CONSUMING MORE ENERGY

In the past the efficiency paradox has been used as an argument against both energy efficiency and conservation. That is certainly not my intention here.

On the contrary, as we face the greatest energy challenge of our times, the need for energy efficiency has never been greater. But at the same time, we must learn from our past experience with the paradox.

If efficiency is to lead to actual conservation, consumers must ultimately be kept from reaping the benefits of those initiatives in the form of ever greater energy consumption. In short, energy prices can't be allowed to fall, or else history has shown that we will just end up consuming more energy.

We must become more energy efficient without the reward of lower energy prices. Then and only then can efficiency lead to real conservation.

But if I'm not going to be rewarded with cheaper energy prices, why should I bother becoming more energy efficient?

The answer is simple. Energy prices, and in particular oil prices, will go up that much more and that much faster if we are not efficient. At the end of the day, efficiency can't allow us to consume any more oil than an already flagging world supply curve can offer. The challenge of oil depletion is to consume less energy, plain and simple. But that's an unpalatable option to most of us. Our lives are too dependent on the huge levels of energy that we consume. Instead we resort to myths that allow us to pretend that we can circumvent the resource constraint and continue to consume ever greater amounts of energy in our lives.

Energy efficiency is one such head fake. It leads us to believe that improvements in energy use are tantamount to resource conservation. But that is only so in an imaginary world where there is no economic growth. And that's not a world that either you or I would want to live in.

But neither would we want to live in a world where the laws of thermodynamics no longer apply. As you may recall from high school physics class, energy cannot be created or lost. When anyone talks about creating energy, what they really mean is that they have found some new way to use energy. Otherwise they are just plain wrong. A good example of developing a new way to harness energy would be a wind turbine or a solar panel.

A good example of being just plain wrong is corn-based ethanol.

THE ETHANOL HEAD FAKE

The second law of thermodynamics tells you that ethanol is not going to solve our energy problems. No system is 100 percent efficient—that's what the second law of thermodynamics says. Energy is always escaping as heat or noise or some other effect. That is, the energy is still there—it's just doing something other than what you want it to do. So while a wind turbine certainly puts to good use the energy of the wind, it also squanders much of it in the heat caused by the friction of the moving parts, in the thumping sound of the giant blades and in many other ways that that are not considered "work." (The grid of high-voltage wires that moves that electricity around also "leaks" a lot of that energy.) The average efficiency for a wind turbine is about 20 percent. In other words, about one-fifth of the wind energy that turns the blades is available to do useful work like charging a BlackBerry or making toast. The rest is lost.

If that sounds like a bad deal, consider that your car engine is about as efficient at turning energy into something useful like getting you to work, and you paid a lot more for that gasoline than you would for the wind. Roughly 80 percent of the energy in that expensive tank of gas doesn't get used to make your car move.

But for a really bad thermodynamic deal, look no further than ethanol. Never mind for a moment that grain alcohol contains less energy than the gasoline it is meant to replace. And briefly set aside the fact that, like gasoline, most of the energy in a tank of ethanol will go to waste. The fact is that ethanol (like hydrogen) is just a way of changing energy from one form to another. And the laws of thermodynamics dictate that much of it is going to be lost on the way. If you start with a fixed amount of energy at one end of the process, you are guaranteed to wind up with less at the other end.

That makes for a losing proposition when it comes to dealing with energy scarcity.

Biofuels are a powerful myth, one that is ultimately much more destructive than the illusion that efficiency will save the day. Biofuels nourish the myths of energy self-sufficiency and of greener power—and lull us into the reassuring belief that things can go on more or less as they have. And there are no more dangerous myths in the entire biofuel paradigm than the ones surrounding America's favorite biofuel—corn-based ethanol.

Ethanol is produced by distilling simple sugar into alcohol. While ethanol can be produced from many biological sources, 95 percent of the ethanol produced in North America is distilled from corn.

Ethanol has long been used as an additive in gasoline, where it can constitute as much as 10 percent of the fuel mix for a standard car. Cars with refitted engines can run on as much as 85 percent ethanol, hence the now popular name "E85" for the fuel in the US Midwest, where it is most commonly sold.

While interest in ethanol has been recently piqued by both soaring oil prices and growing environmental concerns over greenhouse gas emissions, the technology behind its production and use is by no means new. Henry Ford's first car ran on 100 percent ethanol fuel, while the first car to run on a mixture of ethanol and gasoline was built in 1905. What is new is the notion that a widespread substitution of ethanol for gasoline is economically viable.

Ethanol production has been surging around the world, particularly in the United States, where huge increases in national production were mandated by the recent Bush administration in Washington. Production soared from around 1 billion gallons per year in 2000 to over 6 billion gallons by mid-2007.

But a much larger expansion was planned. George W. Bush mandated a national target of 35 billion gallons per year by 2017, almost a sixfold increase from what were already record production levels in 2007. (The European Union's current target is to have 20 percent of its fuel come from crops by 2020.) If that sounds like good news, keep in mind that coming even remotely close to that target for national ethanol production means that your average American family will very soon be eating its last ear of corn.

Ethanol production already gobbles up almost a third of America's total corn crop. If we continue to move toward that 35-billion-gallon target, corn will no longer be available as a food source for either humans or livestock. We will be taking food off the table to feed our cars.

If that doesn't make a whole lot of intuitive economic sense to you, it shouldn't. And it would never happen in a truly free market. But just as oil markets in Saudi Arabia and Venezuela hide the true cost of the energy they are consuming there, so market-distorting US government policies help disguise ethanol's many shortcomings.

The key reason it doesn't make economic sense to convert corn into motor fuel is the huge amount of hydrocarbons that must be burned to first grow and harvest the corn and then to distill the ground cornmeal into ethanol. Like all sources of nonconventional energy, corn-based ethanol suffers from a diminishing energy rate of return. The energy return is so negligible, in fact, that without massive subsidies no one would consider ethanol a viable fuel source.

Corn doesn't just grow by itself. Instead, corn cultivation gobbles up almost 40 percent of all the fertilizer used in the United States. Where do you think all the ammonia that is used in fertilizer comes from? Natural gas. Run out of natural gas, and you run out of ethanol too.

And how does the corn get planted and harvested? Those big tractors and combines run on diesel fuel, not E85. So do those big trucks that move the corn. And the corn doesn't just grind itself into a fine meal and then heat itself in the distillery plants springing up like weeds in the American Midwest. Those processes require energy, and lots of it. For the most part, that energy is supplied by burning natural gas or even, in some cases, coal. These energy inputs are not trivial—a North American farm uses the energy equivalent of between three and four tons of TNT per acre. Modern industrial farming is largely a strategy for turning fossil fuels into food. And if you want to turn that food back into fuel, you can expect the second law of thermodynamics to take its share.

Lastly, ethanol can't be shipped in a pipeline like natural gas or oil. That limits the range of the markets it can serve, since it must instead be transported by truck or rail. Again, those trucks and trains are running on diesel, made from crude oil, not corn. We would have been better off eating the corn and using the billions of dollars of development money and subsidies on something that actually works.

Add it all up, and about three-quarters of the energy in a gallon of corn-based ethanol comes from the combustion of natural gas, diesel and coal used during the various stages of growing corn, transforming it into ethanol and then transporting it.

So why is ethanol production expanding so rapidly?

Just as Canadian bitumen needs to be injected with hydrogen to resemble real crude, American ethanol requires massive injections of taxpayer cash to begin to look like an energy source. Throw enough subsidies, tax breaks and tariff protection at it, and corn-based ethanol can be made to look like an economically and environmentally attractive alternative to imported oil.

Let's start with Washington's Ethanol Excise Tax Credit, which gives a tax break of 45 cents for every gallon of ethanol that is sold as a motor fuel. The subsidies don't stop there. Most major corn-producing states give additional cash both to ethanol processors and to farmers growing corn for ethanol.

And if all of those subsidies aren't enough, then there is the tariff protection given to American farmers and distillers from a crippling 54-cent-per-gallon duty on the much more efficiently produced Brazilian ethanol, made from sugar cane. That duty effectively keeps Brazilian biofuel (which has a much better energy return on investment than corn-based fuel) from ever filling an American gas tank.

As US ethanol production soars, so too does the cost of all those subsidies paid by the American taxpayer. In 2007, the handouts surpassed the $8 billion mark, offsetting as much as half the cost of production. That US ethanol subsidy is almost as much as the $10 billion or so that the government of India spends subsidizing oil—a practice routinely denounced by free-market Americans. And just as India's oil subsidy will become more expensive as world oil prices continue to rise, so too will America's ethanol subsidy as it meets ever-increasing targets for ethanol production. If corn-based ethanol production in America reaches the 35-billion-gallons-per-year target set by former president Bush, US subsidies will balloon to a staggering $25 billion.

And just what exactly are American taxpayers getting in return? While they may not realize it yet, the net energy loss that occurs in growing corn for ethanol production is a huge financial hit paid for in tax dollars: the subsidization of a whole lot of natural gas and diesel fuel that is burned to create a gallon of ethanol fuel.

But if it all seems worth it to you if ethanol is a more environmentally friendly way to fill up for your commute, I'm afraid that is just another head fake. Not only does ethanol not replace gasoline derived from crude, it is no better environmentally, and in some cases much worse. Because so much of the energy in corn-based ethanol comes from burning hydrocarbons in the first place, the cleaner tailpipe emissions turn out to be the tip of a pretty dirty iceberg. If an ethanol distillery is fired by coal rather than natural gas, the fuel that it produces will actually contribute more to changing the climate than regular gasoline. And don't forget all the fertilizer dumped on cornfields. When it is exposed to the weather, the nitrogen in the fertilizer generates nitrous oxide, a greenhouse gas 296 times more damaging to the climate than carbon dioxide.

The United States is certainly not the only country that gets head-faked by fuel alternatives that are more problematic than the fuels they are intended to replace. One study shows that biodiesel derived from palm oil grown in Indonesia is in fact 10 times more damaging to the climate than conventional diesel. Meanwhile, it is expected that the Indonesian rainforest will be 98 percent gone by 2022, cleared away in order to make room to grow this supposedly "green" fuel. Another study calculates that clearing forest to plant biofuel feedstock results in carbon emissions as much as 400 times worse than conventional fuel.

INFLATION AND PEAK CORN

But the biggest knock against ethanol isn't its bogus energy economics or its bogus greenhouse gas emission savings—it's the impact on food prices and inflation in general, which is very real.

While corn-based ethanol will make a negligible difference in meeting energy demand, it has already made a major contribution to higher food inflation.

Diverting an ever-increasing share of the American corn crop from human consumption and livestock feed to energy production has immediate consequences for the price of food. US corn prices soared 60 percent between 2005 and 2007 as ethanol production claimed an ever-increasing share of US corn production. But the impact of soaring domestic ethanol production on food prices is by no means restricted to corn. As ethanol-driven demand for corn pushes up corn prices, what do you think American farmers do?

Like any rational economic agent, they respond to market signals by switching from growing lower-valued crops to higher value-added corn. And thanks to Washington's mandating 7 billion gallons of ethanol production a year, pretty well everything growing on American farms these days takes a back seat to corn.

As more and more acres are converted to the production of corn, fewer and fewer acres are available for other crops that compete for the same land use. Soon the price of these other crops starts to rise as well.

In 2007, corn cultivation in the US rose by 20 percent while the cultivation of all other crops fell by 7 percent. Between 2000 and 2007, corn went from constituting 30 percent of the eight largest crops grown in the US to 40 percent.

The first casualty was soybeans, since corn and soybeans are typically planted in rotation, but the impact has spread to other crops as they too cede land use to corn cultivation. Increases in the price of rice have already sparked riots and hoarding around the world, and the price of wheat more than doubled between mid-2006 and mid-2008.

Rising corn prices are in turn transmitted up through the food chain, given that 50 percent of corn grown in the United States is used for animal feed. Just about any animal products you come across in the butcher section of your supermarket was fed a diet of corn (including that farmed salmon, by the way). The eggs you buy come from corn-eating chickens. The milk and cheese and yogurt come from corn-eating Holsteins. It is not hard to see where this leads: as corn prices soar, so does the price of what we feed our corn to, and ultimately the price of our food itself.

And the butcher and dairy aisles are just the most obvious places to go looking for traces of corn. Literally thousands of other things in the supermarket contain corn in some form, whether it is something as obvious as corn oil or as esoteric as the unidentifiable things that appear on just about every list of ingredients—stuff like lecithin and dextrose and xanthan gum. It all comes from corn. The sweetener in the soft drinks comes from corn. Corn is in the frozen foods and the salad dressings and even the diapers. If peak oil leads to peak corn, we may find that things look different in our shopping carts. But long before then, we are going to find that supper is getting more and more expensive.

Not surprisingly, food inflation is suddenly on the rise both globally and especially in the United States, where the push for corn-based ethanol fuels has been the greatest. Even as the recession has quashed headline inflation to zero, food-price inflation in the US was still running at a red-hot 5 percent at the beginning of 2009.

Washington didn't order the United States to produce 7 billon gallons of ethanol a year without some constituents benefiting from the move. Indeed, corn-based ethanol production provides huge benefits to some, and consequently has its supporters.

Farmland in the American heartland is more valuable than it's been in over thirty years due to soaring crop prices. John Deere dealerships haven't seen farm equipment sales this robust in decades, as once cash-poor and debt-ridden farmers suddenly have a lot of income to spend. Potash producers in Saskatchewan were working overtime to try to keep up with the American market's seemingly insatiable demand for fertilizer despite a quintupling in the price of a ton of the stuff in 2008. For those blessed by ethanol production, times have never been better.

But for energy consumers in the United States as well as for American taxpayers, who are ultimately footing the bill for all those subsidies, the production of corn-based ethanol is a costly head fake whose attraction is all about local politics, not about national energy security, or, for that matter, even energy common sense.

Surging food inflation and mounting budgetary costs are already beginning to undercut public support for ethanol's ill-conceived subsidies. The sooner that public support collapses, the better. The only thing this renewable energy policy will fuel is inflation.

LEARNING THE WRONG LESSONS

A few decades later, it seems that the worst consequence of the oil shocks of the 1970s was not the panic or the lineups at the pumps, not the stagflation or the arrival of the Pinto. Perhaps what will come back to haunt us the most are the lessons economists and policymakers took away from those times—the wrong lessons.

While we routinely pat ourselves on the back for reducing the amount of oil we burn to produce a dollar of GDP, our economies nevertheless continue to ever more efficiently consume more and more oil, making them even more vulnerable to oil prices.

The fact that we can support a larger economy today for a

given level of oil consumption than we could have thirty or forty years ago should be of limited solace to us. The same efficiency paradox that has prevented the average car owner from cutting his fuel bill or the average homeowner from reducing her power bill plays the same role in the economy as a whole. Oil per unit of GDP in the US has fallen over 50 percent since the first OPEC oil shock, but total oil consumption has risen by 20 percent nevertheless.

The OPEC shocks didn't wean us off oil. They just prompted technical change that has made us even more leveraged to the stuff. And when we can no longer expand supply, we risk living in a stagnant world economy that may no longer be able to grow. That world, which could be right around the corner, is going to feature a lot fewer drive-thrus, a lot more bicycles and, no doubt, less Atlantic salmon on our dinner plates. In short, it is going to be a whole lot smaller.

PART
TWO

CHAPTER 4

HEADING FOR THE EXIT LANE

IF YOU HAPPENED TO FIND YOURSELF in Belgrade in the mid-1990s, you already know how expensive gasoline can change a city.

We may have thought that $4-per-gallon gasoline was expensive, but it was still a lot cheaper than mineral water. Thanks to UN sanctions, however, that wasn't the case in Serbia during the series of conflicts that marked the breakup of Yugoslavia. Back then, the only way to put gas in your car was find a black market gas station—a roadside stall selling Evian bottles full of smuggled gasoline. And with that gasoline at 10 German marks a bottle (about $8 US—or roughly $36 a gallon), many people just left their cars at home and found some other way to get where they were going.

So, while the country was at war and convulsed with political tension, the effect in the capital was an eerie calm. Lonely buses and a few taxis plied the empty boulevards. While other European cities were roaring with traffic and choked with the soot from diesel exhaust, Belgrade was nearly silent and the air was fresh and clear. The cafés and bars were full and the sidewalks still bustled with people going about their business, but the parked cars never moved.

That is what a major city looks like without gasoline. The good news is that the smaller world you will soon be living in will feature fresh air and quiet thoroughfares. You can guess what the bad news is.

Nowhere will the adjustment to growing energy scarcity be greater than on the road. Well over half of the oil consumed every day in the United States is consumed as a motor fuel. It's no different in Canada or Australia or any other affluent country where people live far apart from each other and love to drive. Consuming less oil will mean consuming less gasoline and diesel. And that means that one way or another, we are going to have to get off the road.

We will soon be driving less for exactly the same reason the Serbians decided to walk or ride their bicycles: we won't be able to afford to drive the way we have been accustomed to. And we don't have to peer that far into the future to see such a world. Just look what happened in 2008. Gasoline prices in the United States rose from around $1.80 per gallon in 2004 to over $4 per gallon in mid-2008, an increase that now dwarfs even the price hikes motorists had to contend with during the OPEC oil shocks. The run-up since 2002 is almost four times as great as the increase following the Iranian Revolution.

Even in inflation-adjusted terms (or "constant dollars," as economists call them), the price of gasoline climbed to levels greater than when cursing 1970s American motorists had to line up at gas stations due to fuel shortages. Suddenly, the OPEC oil shocks that had seemed like a worst-case scenario don't look so bad. In fact, the process of having our world shrink promises to be much more wrenching this time around, if only because it has got so much bigger since the last energy crisis.

If we measure the cost of driving against other prices in the economy, we will find that it has lagged well behind the general rate of inflation. As driving became cheaper, more people drove. America has gone from being a land where every garage had a car in it to a land where every home now comes with a double garage.

That is because it is a lot cheaper to buy and run a car today than it was thirty years ago. Whereas energy expenditures gobbled up almost 20 percent of the average household's paycheck in 1980, twenty years later that ratio had fallen to 5 percent. And keep in mind that people were driving fully 30 percent more in 2000 than they were at the time of the OPEC shocks.

We have been driving more and paying less to do it. Soon we will be driving less and paying more. Dependence on your car is starting to look like a bad idea.

As much as it stung to fill up your tank when gasoline prices spiked in 2008, you can expect it to hurt even more in the near future. In the next business cycle, pump prices will rise to as high as $7 per gallon as oil becomes even scarcer on world markets. That's 70 percent higher than the peaks we saw in 2008, which sent the world economy crashing into recession.

What will the driving landscape look like at $7-per-gallon pump prices? Europe is as good a place as any to start getting an idea of what American highways will soon become. Thanks to the kind of crippling fuel taxes that would inspire most US motorists to take up arms against their government, European motorists have been paying those pump prices for years.

But European and American drivers are very different. Even the United Kingdom, a country culturally and historically close to North America, provides huge contrasts. Whereas over 90 percent of Americans drive to work, only 60 percent of Brits do. And while 60 percent of American households have a second car, only 30 percent of British households own a second vehicle. While the United States, not surprisingly, has the most cars per capita in the world, Britain comes in a distant twenty-third. Most of the other English-speaking countries, however—Australia, Canada and New Zealand—are fashioned in the American mode; those three come in third, fifth and sixth respectively.

Not only do Americans own more cars than Brits, they drive them more. On average, your typical American uses his car four times a day for a total driving distance of almost 21 miles. The average Brit makes only two driving trips a day, and the total distance traveled, 7 miles, is a third of the total distance traveled by the typical American.

And of course, 30 percent of British households don't even own a car. By comparison, fewer than 10 percent of American households don't own at least one vehicle. Somehow, a third of the citizens of the United Kingdom manage to go about their business without a car.

Brits not only own fewer cars on a per capita basis than Americans, but they drive very different types of cars. That's true even if they are driving a car made by a North American auto manufacturer like Ford or General Motors. Half of the car models that General Motors sold in the UK and the rest of Europe haven't seen the light of day in a North American showroom. Until now, that is.

If North Americans want to see what their next car will look like, they need only check out the showrooms in Frankfurt and London today. Those small, underpowered cars that couldn't crack the North American marketplace are going to be the most popular cars in North America in a world of triple-digit oil prices, just as my Toyota Corolla suddenly stole the limelight back in the 1970s.

One great irony of tomorrow's car market is that it will be the

North American auto producers Ford and what's left of GM and Chrysler as they emerge from bankruptcy protection that will lead the charge toward efficient compacts. The companies that introduced the Hummer to the world, and the 440-cubic-inch muscle-car engine, will soon be trying to sell you a quirky little hatchback. One of their best-kept secrets is that they have been producing these cars for years, and selling them around the world to drivers who care more about efficiency and less about powerful acceleration than Ford and GM think *you* do.

While GM struggled in its home market, in Europe it produced the second-best-selling compact car in a market where just about everybody drives a compact car, behind only the Volkswagen Golf. It is called the Opel Astra. Despite the immense popularity of the Golf over here, GM didn't think North Americans would buy the Astra—until it became clear to them just how wrong they had been in their strategic thinking. Just as gasoline prices started climbing toward $4 per gallon, GM announced that it would begin importing its own Astras from its Belgian plant. Of course, since then bankruptcy forced GM to put its Opel division up for sale but don't be surprised if the new owners, either Canadian parts giant Magna International or China's Basic Automotive Industry (BAIC), start exporting these hitherto unsellable European models to the North American market.

And that will be equally true for Audi or BMW or any other European car manufacturer that sells vehicles over here. Most of the models these firms make for their home market in Europe haven't made it across the Atlantic. Cars made for the European market are either too small or too underpowered (or both) to meet North American driver tastes. North Americans may be surprised to learn that Volkswagen builds a car even smaller than the Golf, called the Polo, but you may soon see one parked on your street. If the phenomenal success of the Mercedes-built Smart car is any indication, European carmakers will be rethinking their model lineups. When North Americans liked their vehicles big and powerful, they could do big and powerful. But if North Americans want small and efficient, they can do that too.

CAN WE GET OFF THE ROAD?

If you have ever tried to negotiate a roundabout in Paris, you already know that Europeans are expert drivers. Somehow they navigate tiny streets and boulevards packed bumper to bumper and mirror to mirror at speeds that make a North American's head spin.

But the secret of the European transport system is not their ability to go out for milk (or, more likely, wine) at rally-car speeds, but rather their recourse to public transit. And while Europe may have smooth highways where family sedans hurtle along at 120 miles per hour, they also have high-speed trains that will get you where you are going at 200 miles per hour. This dense network of subways, trains, buses and trolleys makes owning a car in Europe a handy luxury but hardly a necessity. And as we have seen, those Europeans who do own cars don't drive them nearly as much as we do. They can hop on the metro to go out to dinner, and stroll to the corner when they need to pick up a baguette.

If that sounds appealing, get ready to invest billions in infrastructure.

My city's subway system is already running near full capacity. Trains are running as close to each other as safety margins permit. Toronto's subway and public transit system may not be New York's, but it's no worse than what's found in most American cities these days, and certainly a lot better than what's found in commuter cities like Los Angeles. Yet even the Toronto transit system is stretched to capacity these days.

As drivers across North America, and in Australia and New Zealand as well, decide to become more European and commute by public transit, they are going to find out that their public transit systems are not even close to being ready for the challenges posed by a smaller world. Transit ridership increased by about 5 percent in US cities in 2008 as Americans found a way to drive 83 billion fewer miles over the first three quarters of that year than up to that point in 2007. But how much further can this trend go if the subways are full?

People will not just abandon their cars en masse, but they will use them less and less. If I have to drive to work, I might not be able to afford to use my car anywhere else when the cost of filling the tank is going to run north of $100. And for some 10 million Americans or so, the cost of driving will rise so high that they really will have to get off the road. As people start to park their cars for longer and longer periods, they will increasingly want to get on the subway or LRT. And when they do, the legacy of North America's past transportation choices will come back to haunt the continent.

In the early postwar decades, while Europe was investing heavily in its public transit infrastructure, America poured its wealth into new roads and highways. Aggressively lobbied by Detroit's automakers, both state and federal governments spent billions of dollars of taxpayers' money on building the world's premier road system. And that was, after all, only fitting. America is quintessentially the land of the car.

But it wasn't always that way.

Most people think of Detroit as Motor City. But back in the 1920s, that place that was to become the center of America's auto industry had an interlocking network of buses and electric streetcars that made it one of America's showcase public transit systems. At the end of World War II, the city's public transit system provided over a million rides a day. A decade later, that same vital system was effectively dismantled, displaced by the onslaught of vehicles that were pouring out of its thriving factories. Today, Detroit is the largest city in the United States without some kind of light rail system. What happened?

In the 1920s, things looked a lot like they might in the smaller world of the future. In the United States, 90 percent of all trips were taken on electric-powered rail, and only one in ten people owned a car. Many people thought the auto market was saturated, and GM was losing money. And so the carmakers set about creating new markets. GM built cars and buses, not trolleys, and thus had every reason to lobby hard to see cities across North America abandon electric rail. The process of switching from trolleys to buses began in Detroit in the 1930s but was interrupted by the Second World War, when the need to conserve precious fuel temporarily favored electric-powered streetcars over buses again. But when the war ended, Detroit's fight against commuter rail resumed full force.

GM wasn't the only company with something to gain from the demise of the trolley. Standard Oil of California and Firestone Tire and Rubber also had a lot at stake. The three firms joined forces and created National City Lines in 1936, which itself gave birth to another subsidiary, Pacific City Lines, in 1938, and to yet another company in 1943, called American City Lines. Their purpose was to buy and tear up rail lines, which they did with remarkable success. They targeted over a thousand municipal electric rail systems, and managed to motorize 90 percent of them through a combination of pressure tactics, outright purchasing and destruction of private rail systems, and bribery of the officials in charge of the public system. (In some cases, Cadillacs were offered to officials as incentive to switch their systems to buses.) A similar process unfolded north of the border, in Canada, although somehow a few streetcar lines have survived to this day in Toronto.

If this sounds like some giant corporate conspiracy to undermine public transit, a US federal grand jury would agree. GM was indicted for criminal conspiracy in 1949 for its role in running the Los Angeles rail system into the ground, but that did little to slow the company down. In 1974, the auto industry's role in gutting America's light rail system was investigated by the Senate Subcommittee on Antitrust and Monopoly. But by then, there was really no light rail left. Detroit's streetcars had been sold to Mexico City for a song, but at least the Mexicans got good use out of them. Supposedly so antiquated that they could no longer do the job in Motor City, the streetcars performed yeoman service in Mexico until 1985, when they were destroyed by an earthquake.

"What's good for General Motors is good for America" was the adage of the golden age of American motoring. And most of all, what General Motors wanted was for Americans to drive. GM's dream of ever-soaring car sales meant moving people farther and farther into the suburbs. With the convenience of the car, and with roads to drive it on, the distance between where people worked and where they lived kept on increasing.

That meant linking cities and states all across the nation through a complex labyrinth of interstate highways. A committee appointed by President Eisenhower and chaired by Lucius D. Clay, who also just happened to be on GM's board of directors, recommended a massive overhaul of the American highway system. The Interstate Highway Act was duly passed in 1956. It called for a public works project on a historical scale: 41,000 miles of new highways linking every city of more than 50,000 inhabitants, and the widening of countless existing roads.

As the distance between work and home grew, city boundaries gobbled up more and more of the surrounding hinterland. Farmers sold out to developers, who then bulldozed and paved over the pastures and fields to build new communities for the millions of people pouring out of the cities in their shiny new automobiles—virtually all of which were manufactured in North America. And the more extensive the roadways, the farther those suburbs could push into farmland and forests.

Americans, for the most part, were only too happy to follow the trail that Detroit was blazing for them. The relatively cheap real estate prices of suburbia meant that you could buy double, sometimes even triple the home or lot that you would be able to afford back in the city.

It would mean not only a bigger home, but also a sprawling backyard for the kids to play in, instead of the crammed city lots and traffic-congested inner-city streets. And since all the houses were brand spanking new, it meant that they would come with all the latest appliances and finishes.

For that matter, not only were the houses brand spanking new, so was the entire infrastructure. Brand-new schools, brand-new sewer and hydro systems, brand-new hospitals were all part of the suburban package that by the 1950s had come to symbolize the American dream in bricks and asphalt. Cheap energy has allowed it to remain so for the last five decades—only now those postwar suburbs seem like the inner cities compared to the far-flung rings of suburbs and exurbs of monster homes with tiny saplings planted in the plush new sod out front.

The one thing missing from that dream was public transit. Why build public transit in suburbia when everyone has a car?

After all, if folks didn't have cars, they wouldn't be moving to the suburbs in the first place.

Public transit may have seemed like a frivolous expense at the time when the suburbs were built, but its omission in a world of triple-digit oil prices is a fatal flaw.

In the new world of triple-digit oil prices, North Americans will have to become more European in their driving habits. But we can impose European behavior only on those drivers who have the European alternative of taking public transportation, whether they live in the United States, Canada, Australia or New Zealand—or Greenland, for that matter. But when it comes to taking the bus, streetcar or subway, America ranks the lowest among all OECD countries. And that is because in many cases there is no transit to take, even if Americans wanted to take it.

As a first approximation of that potential subset of American drivers that could adapt to European commuting habits, look at the roughly 57 million American motorists who both own at least one vehicle and have broad access to public transit. That's generally defined as having no more than a half-hour total door-to-door walking time from house to work. Not that many of us would consider walking that far.

Those 57 million potential "Europeans" are slightly more than half of the total number of households in America that own at least one vehicle. These are the people who will be able to commute like Europeans when oil prices dictate that they do.

Now, in Western Europe, two of every ten households don't own a car. So if the same percentage of our 57 million Americans capable of emulating Europeans actually do so, as fuel prices will certainly entice them to, we can expect some 10 to 12 million American households to literally get off the road.

Those who do are most likely to be low-income drivers,

everything else being equal. You and your next-door neighbor may prefer an expensive commute to a crowded one. It is not as though Londoners don't experience rush-hour commuter traffic just as congested as anything in New York or Toronto. The very wealthy will be the last to park their cars. But what *is* different about North America from most other places is that even the poor drive.

In fact, many drive more than one vehicle. There may be more Maseratis or Ferraris on the road in the United States than anywhere else, but what speaks the loudest about the importance of the car in American culture and life is not the driving habits of the rich but rather the driving habits of the country's poor. It is in this income bracket where the rubber really hits the road. Some 24 million American households with annual family income of less than $25,000 own at least one vehicle. Incredibly, there are more than 10 million such households that own and drive more than one car.

Soon they won't be driving any. At least one in five of those low-income drivers will have to give up their second car, which is most likely a gas-guzzling near-wreck. But many will have to give up their first car too, which is likely newer and hence more fuel-efficient.

Gasoline already overtook grocery spending for low-income families in 2008, when prices surged to $4 a gallon, and it will do the same for even average-income Americans in the future. America may be the land of the car, but when faced with the choice of feeding your stomach or filling your gas tank, your stomach is usually going to win. While Americans have steadily bought less gasoline over the decade, the reduction in quantity, at least until the current recession kicked in, had not kept up with the increase in price. Hence, even with chastened driving habits, gasoline sales had grown five times as much as the rest of retail spending in the United States between 2004 and 2008.

Income will obviously play a huge role in determining who stays on the road and who gets off, but where you live will also count. While 75 percent of all Americans living in cities have access to some form of public transit, only 50 percent of American households living in the suburbs have similar access. In rural areas, access to some form of public transit plummets to about 25 percent.

Adaptation to high prices will not be easy for those of us who depend on our cars for every carton of milk and all our errands for our kids. Subways just don't work in the low population densities of the suburbs—in fact, even bus routes function poorly where houses are far apart and amenities nearly nonexistent.

But even those among us who are relatively well prepared for high oil prices will find that their worlds are going to get smaller. Although Europeans drive less than North Americans, it is not as though they don't drive at all. The same goes for city-dwellers all over the world—people in Paris and Tokyo, like New Yorkers and Torontonians, may walk or take the subway to work, but most of them still have cars parked somewhere, and those cars come in handy when it comes time to buy groceries or take the kids somewhere or just head out to the countryside. Try any of that on public transit just about anywhere in the world and you will wish you were back in your car.

Yet the arrival of $7 gasoline over the next several years will see energy spending easily surpass the peak level of past oil shocks, including the one we scraped through in 2008. As it does, cars will be abandoned all across America to become rusting monuments to a past age of cheap fuel.

THE SCRAP HEAP

I drive a nine-year-old Audi, so it's the repair bills, not gasoline prices, that will drive me off the road. But for a lot of us, filling up each week is going to force us to reconsider how we get to work and back, and that is going to mean fewer cars on the road.

How many fewer? Turn the clock back. Turn it right back to 1982, because that's the last time annual vehicle sales punched in below 10 million units per year in the United States. But Detroit's biggest challenge is not that unit sales will have fallen below 10 million units in a recession, though that is bad enough. The real problem is that unit sales are unlikely to recover when the recession is over. While the economy will eventually improve, that improvement will be a double-edged sword for auto producers, since it will bring even higher pump prices—higher than we have yet seen.

Only three years ago, nearly 17 million vehicles were sold in the US. Today's market is little over half that and tomorrow's may be even smaller. Detroit's biggest problem isn't that it produces the wrong type of vehicles, despite the federal government's insistence that any multi-billion-dollar bailout package be earmarked for retooling to build energy-efficient cars. It doesn't just build the wrong kind of cars—it builds too many. There will be fewer and fewer cars on the road in America, and that means Detroit will be producing far fewer vehicles than it was planning to.

Here's why. Just as we can measure the decline in oil reserves, so we can keep track of how fast cars disappear from the road. It's called the "scrappage rate," and it indicates the percentage of existing vehicles that are retired from service each year.

Currently, the scrappage rate is running around 5 percent, which takes some 13 million motor vehicles off the road every year in America. That means 13 million new motor vehicles must be sold to keep the total number of vehicles on the road constant. But we know from history that the scrappage rate itself moves up with oil prices. This reflects the fact that older cars typically average much poorer fuel economy than newer cars and thus become increasingly expensive to run as pump prices soar. The higher the pump price, the more likely people are to retire their gas-guzzlers.

Even a single percentage-point rise in the scrappage rate would mean taking some 14 million aging vehicles off the road every year. If only 10 million new vehicles drive off the dealership lots in that same year, there will be 4 million fewer vehicles on the road. Run that model over a decade and you will start to notice how the drive home in rush hour is getting shorter and shorter—providing, of course, you are not one of the 40 million former drivers now taking the bus.

Three or four years from now, the once super-hot light truck category will have fallen from 60 percent of new vehicle sales in 2006 to less than 25 percent, reversing, almost overnight, nearly two decades of uninterrupted gains in market share. Chrysler is particularly vulnerable to rising prices, depending as it does on light trucks for nearly 70 percent of its profit. Not only will fuel costs make it impractical to drive these vehicles, but in a time of global oil scarcity, they and their owners will soon come to be socially ostracized as the fuel pigs that they truly are. Today, you can't give these gas-guzzlers away. Car lots can't even provide financing for them.

In fact, plunging lease values on returning SUVs have crushed Detroit with losses in recent quarters. Almost a quarter of Ford's stunning $8.37 billion single-quarter loss in 2008 was due to lease losses. Some super-sized gas-guzzlers like General Motor's Yukon or Toyota's Sequoia lost as much as 70 percent of their original value when their three-year leases expired in 2008.

So great are the losses in leasing from plunging vehicle values that Chrysler has already announced that it is getting out of the leasing business altogether, while Ford and General Motors are dramatically scaling back leasing operations and withdrawing leasing altogether for most light trucks.

But it turns out that leasing losses were the least of their worries, as both companies were forced to seek protection from creditors under Chapter 11 bankruptcy. Nearly 50 percent of the North American auto industry's capacity is shut down and it is looking more and more as though much of it will never come back.

But the US is not the only place where car sales are collapsing. In the UK, new car registrations were down 22 percent in February 2009 compared to the February before, while in Australia registrations were down about as much. Sales slumped 28 percent in Canada. And what about America's rivals in the auto sector? In Japan, where production lines are slowing and workers being sent home, sales are down 32 percent, to 35-year lows. German car sales are the worst since reunification— down 14 percent in January. Still, the US has been hit particularly hard. Sales in February fell by 41 percent compared to 2008.

Oil shocks have always turned the rugged capitalists of Detroit into big-time Keynesians. Following the second OPEC oil shock, Lee Iacocca went begging to Washington to save Chrysler from bankruptcy. Then in 2008, taxpayers again heard Detroit's all-too-familiar refrain, "Brother, can you spare a dime?" And Congress voted 370–58 to approve a $25 billion bailout package for the auto industry. Not to be outdone, the European carmakers turned right around and asked for €40 billion from the European

Union to level the playing field, and manufacturers in Canada set their caps out begging for billions of their own and talked openly of moving thousands of jobs south of the border if they didn't get what they wanted.

The stakes are high as all of the big three in Detroit are rapidly running out of cash. At the end of 2008, Congress threw GM and Chrysler another $17 billion lifeline to avert an immediate collapse of both companies and their production lines throughout America. And Canada ponied up $4 billion of its own, to protect its share of GM and Chrysler jobs. But still, the production lines grind a little slower every day, and each drop in production reverberates back through the auto-part supplier industries to the rest of the economy.

Facing the collapse of GM, both the American and Canadian governments were ultimately forced into buying direct equity ownership in the bankrupt auto giant. A $50 billion infusion of American taxpayer money bought Washington a 60 percent share in the company while a $9.5 billion infusion of Canadian taxpayer money bought Ottawa and the provincial government of Ontario just under 12 percent.

But there are, no doubt, a few Detroit auto executives secretly taking solace from others' misfortunes. In a sobering display of the efficiency paradox at work—and also of the shortness of memory we sometimes fall victim to—consumers are responding to falling prices at the pumps by rejecting the very type of vehicle Washington wants Motor City to build. Sales of the Toyota Prius are down around 30 percent. Just as the economics of a Canadian tar-sands operation don't make sense when oil is cheap, neither do the economics of driving a hybrid.

Just as recession-priced gasoline makes car buyers wonder why they should give up their lumbering SUVs in favor of sleek, fuel-efficient hybrids, it makes auto executives wonder why they should bother retooling their assembly lines. If people want trucks, why not build more trucks?

Because all those SUV and light truck plants make sense only as long as oil prices stay under $50 per barrel. It's the triple-digit prices that killed their sales. Since any assistance from the federal government is tied to commitments to build more fuel-efficient vehicles, Congress is obviously telling Detroit it believes oil prices will rise. It may also be telling Detroit that the price for carbon emissions is going to rise as well. But at the end of the day, while Detroit is more than happy to take Washington's money, it will be consumers who decide what type of vehicles auto companies will build.

And one can certainly understand if consumers are more than a tad confused about what type of vehicle to buy. Within three months of the $4-per-gallon summertime gas, the deepest recession in decades has hammered pump prices to less than half of that. If you're one of the few Americans who will be actually buying a new vehicle in 2009, what do you buy? Your oil price forecast will pretty well determine your vehicle choice. You can afford to buy an SUV at recession-priced oil, but if pump prices recover back to what drivers were faced with over the Memorial Day weekend of 2008, you will be buying a Prius instead.

If you are thinking of buying a car in the next little while, the bottom line runs pretty well like this: Oil is cheap only when we can't afford to consume it — in other words, during a recession. So the only time we can afford to fill up our vehicles is when we are losing our jobs and can't afford to buy one in the first place. When the economy improves, and we have the financial wherewithal to buy a vehicle, gasoline prices will have risen back to the point where we will once again no longer be able to afford to drive them. Either way they are toast.

Unless, of course, our next cars don't run on gasoline. But what are the prospects of that?

THE ELECTRIC CAR

Karl Marx said that history repeats itself, first as tragedy, then as farce. We've seen both acts when it comes to the electric car.

We hear a lot about how hybrids and plug-in electrics are the vehicles of the future, but we could all be driving gasoline-free (and greenhouse-gas-free) cars today if it were not for a tragic wrong turn about a century ago. Back in the nineteenth century, Victorian ladies and gentlemen were whizzing around London dodging horse-drawn carriages in their electric cars. That's right. The technology we are pinning our hopes on today was already up and running when men still wore swords when they went out on the town.

Though today's car manufacturers and regulatory bodies worry that battery-powered cars may be too slow, an electric car was clocked at over 100 kilometers per hour (that's 62 miles an hour) in 1899. And while we keep hearing that current battery technology just can't provide electric cars with sufficient range, a Detroit Electric managed to go 211 miles on a single charge back in the days when city trolleys were pulled by horses.

Around the time that Ford's first Model T was rolling off the assembly line, electric cars were more popular than internal combustion models, and it is not hard to see why. They were quiet and clean and reliable, particularly in comparison to the clattering and often dangerous gasoline-powered cars they were competing against for market share. Those wheezing, rattling gas-guzzlers tended to be driven not by their owners, but by chauffeurs who doubled as an early version of today's roadside assistance—if you wanted to get around, you needed someone to start the thing for you and to be standing by to repair it when it inevitably broke down. Meanwhile, electric cars were marketed specifically to people like physicians who needed to be able to get into their cars and get where they needed to go without waiting for someone to get the machine running. (Physicians actually made house calls in those days.)

Then came what now looks like the tragic chapter in the history of the automobile. In 1913, Cadillac developed the first electric starter to replace the cumbersome crankshaft starters we have all seen in old black-and-white footage. Now, for the first time, people without the physique of a weightlifter could easily start their own cars. With that, as well as the steadily falling price of a mass-produced car like the Model T (the sticker price fell from $850 in 1909 to $300 in the 1920s—about $3,200 in today's dollars), the internal-combustion-driven vehicle quickly took off in a cloud of smoke. All that remained from the electric car's earlier heyday were electric-powered forklifts and golf carts, and electric streetcars and trolley buses, which most cities have phased out.

Then, as if to prove Karl Marx right, the decision makers in Detroit decided to reenact this chapter in automotive history as farce.

Back when SUVs were beginning their rise to dominance in the North American market, the California legislature decided that it was time to do something about vehicle emissions, and brought in the Zero Emission Vehicle (ZEV) mandate, which required the biggest car manufacturers selling cars in California to ensure that by 1998, 2 percent of the cars leaving their lots had no tailpipes. That figure was to rise to 10 percent by 2003.

Faced with legislation that would force them to change the

way they did business, the carmakers did what they do best. They marshaled their legions of lobbyists and mobilized them to fight the legislation. Just like the CAFE standard was eventually watered down to the point that it became completely useless, the ZEV mandate was eventually repealed.

But not before a number of carmakers did what they were arguing could not be done. They went and built electric vehicles. Toyota produced a battery-powered version of its RAV4 mini-SUV, Ford came out with an electric pickup truck and just about every other company wheeled out something it could boast about too. But the real star of the show was GM's EV1. It was sleek and fast, and the people lucky enough to get leases quickly fell in love with their cars. Just as the hybrid Toyota Prius is now a status symbol among celebrities, the EV1 was proudly driven by stars like Tom Hanks and Mel Gibson.

GM had a real problem. Their innovative new car was a success.

Innovations are rarely commercialized when the new idea blows up the innovator's bottom line. The last thing GM would have wanted to do was get out of the high-margin gas-guzzler business. Over three-quarters of its earnings came from making SUVs, vans and light trucks. It certainly didn't want to be building EV1s instead of SUVs. And it didn't want to build low-maintenance electric cars when it had a good thing going with people who bought their gasoline-powered cars bringing them back to the dealerships for oil changes and other routine check-ups that padded the bottom line. And besides, if the EV1 worked in California, why wouldn't every state and province start demanding that the carmakers build zero-emission vehicles?

What followed was farce.

The company went out of its way to let people know how impractical its car was, how pitifully inadequate its range was. It was almost as if GM was pleading with the public not to buy its EV1.

In the meantime, all of the parties with a stake in keeping the internal combustion engine running got into the game. Aware that the one perceived flaw of the electric vehicle was (and still is) the range imposed by the state of battery technology, the oil companies bought up battery patents to ensure that innovation never made it to market. And the car companies had other tricks up their sleeves too. Faced with the choice between adapting to the newly configured marketplace and fighting it in court, they opted to go to court.

The basis of the lawsuit brought by GM and DaimlerChrysler was that the ZEV mandate violated California's Environmental Quality Act and the federal Clean Air Act. In other words, GM claimed to be defending the laws of California against the state's own policymakers. The final irony was that the carmakers argued that producing zero-emission vehicles would actually increase pollution by making cars more expensive on average and thereby encouraging drivers to hold onto their clunkers a little longer.

A coalition of EV1 drivers and environmental groups like the Union of Concerned Scientists and the American Lung Association saw things rather differently, and they fought the lawsuit with the resources at their disposal. But the farce played itself out all the same.

Big auto and big oil won: ZEV was repealed in 2003, and the EV1 program was shut down. The cars themselves were rounded up and crushed, despite the pleas of their drivers. When the final seventy-eight cars were waiting in a compound to be taken to the crusher, activists raised $1.9 million to buy them from GM, but the company refused.

Now, only a few years later, those EV1 engineers are working around the clock trying to get the Chevy Volt up and running. That is what ever-rising oil prices will do to a company set up to sell cars and trucks designed to run on cheap gas. Now that the SUV market is gone forever, GM is scrambling to get its electric program back up to speed and has staked all its credibility on an ambitious effort to compress development time and get to market before its chief competitors — Toyota in particular.

Despite its massive reorganisation, and the sale and closing of a number of the company's divisions, General Motors has announced that it is still on target to produce the Volt by late 2010. The car will come equipped with a plug-in electric engine that can be recharged at a regular electrical outlet. Maximum range for the electric engine will be about 34 miles, but the vehicle will also come equipped with a small four-cylinder internal combustion engine that will run a generator rather than the drive train. If you want to travel distances beyond the range of the electric motor, the gasoline engine will kick in and recharge the batteries. The combined electric–internal combustion system is expected to get about 100 miles to the gallon in fuel economy. It seems like only yesterday the same company said that 27 miles per gallon was impossible. (Of course, in 1974 a modified Opel T-1 managed to get an astonishing 377 miles per gallon, so the bar is still pretty high.)

But while the potential fuel economy savings are attractive and the technology looks ambitious, the production numbers are insignificant. Only 10,000 Volts will be made in the first year of production. Even when the assembly line is fully ramped up, GM has estimated that only 54,000 to 60,000 will be built each year.

Even if GM hits its production targets for the Volt, that's hardly going to replace the number of gasoline-driven cars that higher pump prices will ultimately take off the road. At an estimated price of $35,000–$50,000, few of the drivers taking the exit lane from American's highways will be able to afford GM's answer to oil depletion. Chrysler is just trying to get into the game with its battery-powered Envi lineup, which is supposed to be on the market between 2011 and 2013. But, will Chrysler even exist then, let alone pioneer an electric car?

Yet, while Detroit is getting its act together, others are already building the so-called car of future. The bestselling electric car in the world, the G-Wiz hatchback, is made by the Indian company REVA. It has been in production since 2001 and is already being exported to the UK, as the G-Wiz i. The Norwegian TH!NK line of cars is already on the road in Scandinavia, capable of highway speeds, and is expected to sell in the US market for as little as $15,000. In North America, the Canadian ZENN car has gained regulatory approval in the provinces of British Columbia and Quebec, along with forty-six US states. And if quirky little hatchbacks are not your thing, perhaps you will find the American Tesla Roadster more to your liking. It does zero to 100 in under four seconds and has a body designed by Lotus. And it costs less than $0.02 per mile to run it. Of course, the base price for a 2009 model is $109,000.

Neither General Motors, Ford nor Chrysler has yet to demonstrate that it can produce a hybrid car profitably. If all three can't, how long will the market, or the taxpayer in GM's case, let them continue to try? In fact, other than the possible exception of Toyota, no company in the global auto industry is making any money producing hybrids, and even Toyota's margins are reportedly paper-thin. In fact, the venerable Japanese carmaker announced in 2008 that it would not make a profit for the first time in seventy years.

Right now there are approximately 70,000 electric cars driving the roads of America out of a total vehicle population of around 247 million. That means that one out of every 3,000 vehicles on the road will be immune to higher oil prices. All others will soon have to pay triple-digit oil prices at the pumps.

Of course, there is no point redesigning car engines to run on an alternative fuel source if there isn't going to be enough of that new fuel source to do the job. The point is not to replace the cars themselves, but to replace the billions of barrels of oil that make them go. It is fine to say that we will run them on electricity, but how are we going to generate that electricity? Are we trying to fit a square peg into a round hole?

The 13 million barrels of oil burned every day in the North American vehicle fleet contain a lot of energy. If we want to charge up those cars with electricity, we are going to need as much power each day as nearly 2 million North American homes would use in a year. Yes, electric motors are much more efficient than internal combustion engines, but the power plant that generates the electricity is not *that* much more efficient. Or necessarily cleaner when it comes to greenhouse gas emissions.

Energy is not so easy to come by that we can simply decide to plug in our cars and go about business as usual. For one thing, we would need to use a lot more electricity than we do today. If expanding our capacity to replace all that oil doesn't sound like much of a challenge, consider what happened on the afternoon of August 14, 2003. That day, the electrical grid that covers the area encompassing New York, Detroit and Toronto collapsed under the strain of the peak summertime load, with disastrous results. I know, because I had to walk down eighteen flights of stairs that day. Elevators, bank machines and telephones all stopped working largely because it was a hot day and there was one too many air-conditioners plugged in. That is how close we are to peak capacity.

Since then, the utilities have done their best to promote conservation, including resorting to bribing customers not to use their air-conditioners and dishwashers at times of peak consumption. Even the lighting in my building has been dimmed discernibly. In the underground parking lot, it has been turned almost off until motion detectors turn it back on when someone walks to their car. And after 5:30 in the evening, well over half of the lights on the ground-floor lobby are turned off. Even the air-conditioning in the summer now seems pretty lame in my state-of-the-art skyscraper, which proudly advertises in its lobby that it will be cutting its carbon emissions 20 percent by 2012. But even with all these energy-saving measures, the fact is that we keep using more and more electricity, and billions will have to be invested in power generation and transmission infrastructure just to meet projected demand and replace aging equipment. Doubling capacity to accommodate the power demands of potentially millions of electric cars becomes a staggering undertaking.

Even if Detroit were producing enough electric cars to make a discernible difference in the amount of oil we consume, we wouldn't have the spare power capacity to charge them up. We can barely keep our air-conditioners humming. If in this part of North America we lived in a climate where we had to run our air-conditioners all year long, we would probably experience the same number of blackouts and brownouts as they do in Kuwait and the rest of the Middle East.

Now, let's say we decide to go ahead and build all these new electric power generation plants. What do we run them on? We have seen that even current consumption rates are drawing down our natural gas supplies at an alarming rate. Before the recession, North American natural gas prices had doubled over the previous five years, making gas no longer a source of cheap power. And in gas-poor Europe and Asia, the fuel is over double the cost in North America.

There is, at least in theory, the option of building hundreds of new nuclear plants to generate the power. But the fact that no new nuclear plants have been built in North America in a generation is not an accident. While nuclear power plants are actually safer than they are usually given credit for, they are expensive to build and to insure, they are unreliable (about half of the UK's nuclear capacity is offline at any given time) and they generate toxic waste that we still have no idea what to do with (though terrorists have very clear ideas about what *they* would like to do with it). Moreover, nuclear plants take on average seventeen years to put in place.

Finland's Olkiluoto 3 reactor is the latest in a long series of industry disappointments. It was meant to herald a global nuclear renaissance, but the all-too-familiar pattern of technical problems, cost overruns, lengthy delays and procedural problems has plagued the new reactor from day one. Ever-more-expensive natural gas and new concern about carbon emissions may make nukes more attractive over time, but the very reasons that made widescale nuclear power problematic in the past remain with us today.

And what about renewables? The idea of a fleet of electric cars running on electricity from wind turbines and solar panels is every environmentalist's dream. But is it realistic on any mass commercial scale? While renewables get huge press profile and just as big government assistance, they account for a only a very small percentage of total power generation. The government of Ontario, for example, hopes to get to 10 percent by 2010. A windpower leader like Germany is at about 15 percent renewable energy today, and Denmark is shooting for 20 percent by 2011. But many experts think this is about as green as you can get. If the wind stops blowing or the sun stops shining, the grid collapses and the lights go out.

The Middle Ages were very green, but is it really an epoch you would want to live in?

In spite of generous government subsidies, renewables supply only a miniscule amount of the energy we use, and even those fractional amounts are available only in the world's richest countries. We aren't going to replace our most important fuel with an energy source that currently couldn't even power our hair dryers if they were all going at once. We would need to multiply renewable capacity many times over before it would make even a noticeable impression on most countries' total energy supply, let alone become the new power source that will facilitate a mass conversion to electric cars.

Of course, there is always the traditional standby—good old king coal—to fuel all the power stations that will be needed to charge the world's fleet of electric cars. Coal is the cheapest, most obvious source of more power. But the emissions from the power plants would offset all the emission savings from switching to electric cars—unless we first develop and deploy carbon capture and storage (CCS) technology (so-called "clean coal") on a scale far beyond anything currently in operation. Without CCS, recourse to our still abundant coal reserves would be flooring the accelerator on global greenhouse gas emissions just as global climate change is telling us to hit the brakes.

The project of building hundreds of coal-fired plants capable of capturing their carbon emissions, and the pipelines to carry all that carbon to suitable sites for storing it underground, would be a monumental undertaking. To give a sense of how much carbon we are talking about capturing and storing, American carbon dioxide emissions tip the scales at over 6 billion tons annually. That is the same mass as more than 76 million Abrams tanks. Imagine trying to find a place to hide 76 million tanks every year, and you begin to get an idea of the challenge posed by storing our carbon emissions. It's not impossible, but it is not going to be easy, and it is certainly not going to be cheap. We will need to see prices for carbon emissions at least in the $60- to $70-per-ton range before anyone is going to even look commercially at carbon capture and sequestration. That's $60 to $70 dollars more than carbon emissions currently cost in the US economy, the Canadian economy or the Australian economy. In short, without putting a very substantial price on carbon emissions in our economy, CCS is nothing more than a pipe dream.

And speaking of monumental challenges, there is always the distant dream of tomorrow's hydrogen economy. But if a fleet of electric cars is difficult to imagine, consider the obstacles to switching to one powered by hydrogen. For one thing, hydrogen is a lot like electricity in that it is not an energy source—it is an energy carrier. Hydrogen does not come gushing out of the ground: you have to produce it, and that takes fuel. Producing sufficient hydrogen to run the US vehicle fleet—about enough to fill 13,000 Hindenburg dirigibles every day—would require doubling electrical generating capacity. And if you want to use renewables rather than coal, gas or nuclear, you would need to cover an area the size of Massachusetts with solar panels or an area the size of New York state with wind turbines. Now consider the problem of building a network of hydrogen filling stations, then the problem of supplying them with tanker trucks (hydrogen can't be transported by pipeline because of the risk of leaks). Then consider that hydrogen is much more expensive than gasoline, and that hydrogen cars cost about a million dollars each. Suddenly that pair of running shoes or that bike has never made more sense. The idea that we can simply switch from gasoline to hydrogen is not unlike Marie Antoinette's suggestion that the starving French peasantry switch from bread to cake.

The challenges of reinventing the car are not insuperable, but they are certainly huge. Will today's lithium-ion battery be able to stand up to years of driving? So far, the one in my wife's laptop couldn't even handle her daily email before catching fire and frying her hard drive. No electric-car maker has yet got around the fact that gasoline is in the end not a bad way to carry energy around—it packs about twenty times as much energy per pound as the most high-tech battery. But don't expect the world's engineers to give up. Triple-digit oil prices will have a way of focusing their attention on oil-saving technological change. Still, how quickly change will come, and at what price, remains to be seen.

In the meantime, more and more drivers will exit from the road as Detroit adjusts to a market that will be little more than half the size of what it has known. As it does, a new administration in Washington embarking on a massive round of infrastructure spending to stimulate the economy might consider the implications of this prognosis. Why earmark billions of taxpayer dollars for new road construction when there will be fewer and fewer drivers using those roads in the future? And why spend even more billions on propping up an auto industry that will soon be doomed to obsolescence by triple-digit oil prices? It's public transit infrastructure, not more cars and beltways, that America (and the rest of the world) will soon need.

CHAPTER 5

COMING HOME

IF YOU HAVE EVER TAKEN A TAXI, you understand the principle that is going to throw the eighteen-wheeler of globalization into reverse.

No matter how far you have to go, you are watching the meter and wondering if you have enough money in your pocket to cover the fare. Get in a cab in Seoul, Korea, and you can probably afford to drive around all day and take in the sights. But if you are faced with the prospect of a taxi ride in London, you might just decide to walk.

Distance costs money.

That will be the mantra of the new local economy.

Take a look at the things you buy at Home Depot or Wal-Mart or the dollar store down the street. Chances are they weren't made anywhere near where you live. Your jeans, your coffee maker, your cellphone, whatever—it was probably made on the other side of the world. But just as they did over thirty years ago during the oil shocks, soaring energy prices will once again have a huge impact on transport costs all around the world. Whether you are moving goods by ship, train, truck or air, the fuel costs of transportation, not the cost of tariffs, are about to become the largest barrier to global trade today.

In effect, soaring oil prices will soon put us in a time machine heading back to a world trade environment that will be broadly comparable to what the world looked like three or four decades ago. Back then, countries traded with each other a lot less than they do today. Their national economies lay safely cocooned within huge tariff walls that protected them from foreign competition.

That's a very different world from the one our economy is currently plugged into. World trade in the days of trade barriers grew at a fraction of recent rates. Fewer goods were exported, and fewer goods were imported. Hence, the domestic market was much more important than it is today. Those were the days when your fishing reel was most likely made by Penn instead of Dawai, when you drove a GM or Ford product instead of a Toyota or Honda, and when even your socks were made in Vermont instead of Indonesia. And in virtually every case where tariffs or quotas compelled you to buy made-at–home goods, you invariably paid more for whatever you bought than you would have if imports had been allowed to enter your local shopping plaza without punitive duties.

Moreover, when countries were exporting or importing back then, they weren't trading with the countries they deal with today. World trade patterns were far more regional: countries tended to trade with their neighbors, not with countries on the far side of the world. There were exceptions, of course. We have always got our tea from China and India. But we used to get our widgets much closer to home. In fact, most of our widgets were made right at home.

That has been changing steadily over the lifetimes of most people reading this book. Between 1960 and 1973, exports as a share of world GDP rose by over 50 percent. And that has made the world better off. Economists have known for a long time that international trade benefits both importer and exporter, and that is why the second half of the last century saw such aggressive efforts to bring down trade barriers. Decades of multilateral trade negotiations have radically changed the world. The emergence of India and China as economic powers can be traced in large measure to the reductions in tariffs and nontariff barriers on the part of the wealthier importing nations. The explosive growth in international trade over the last quarter century is synonymous with the broad globalization of our economies.

Globalization comes with a lot of verbiage these days, but it is just a fancy word for a very simple process: moving your factory to the cheapest labor market in the world. Even better, getting out of the factory business altogether and just buying whatever your factory used to produce from someone else's factory at a fraction of the cost. But before you can do either, you first have to dismantle the trade barriers that used to tie your factory geographically to the market that it served.

Thirty or so years ago, the Chinese wage advantage was a lot greater than it is today. But you didn't see a lot of things imported from China back then. No one was going to stop you from moving your factory there, but trade barriers made damn sure it was hard to sell anything you might have produced in your Chinese sweatshop in any of the major industrialized economies of the world. If double-digit tariff rates didn't do the trick, then ironclad import quotas on everything from clothing to steel kept you and your Chinese factory at bay.

Back in those days, if you said goodbye to your high-wage domestic labor force, you also had to say goodbye to your bread-and-butter domestic market. That market was reserved for homegrown producers employing local, albeit expensive, labor.

That was the quid pro quo: higher prices and hence less purchasing power in exchange for local jobs.

Now all those trade barriers have been removed and firms are footloose and fancy free to move to wherever they can make the most money. As more and more of the world's factories were uprooted from North America and Europe and sent overseas to cheap Asian labor markets, unions back home lost their bargaining power.

When jobs crossed the ocean, wage concessions quickly replaced wage gains in new collective bargaining agreements in once highly protected and highly paying industries like autos and steel. With labor costs typically accounting for as much as two-thirds of the cost of a typical manufactured good, containing wages meant controlling prices.

That's ultimately why the US supported globalization and agreed to give up all those jobs to China. Globalization changes the quid pro quo to lower prices for fewer jobs. At the end of the day, we supported globalization because we wanted to be able to buy cheaper computers, cheaper vehicles, cheaper clothes and cheaper furniture. Wal-Mart parking lots were jammed with North American workers buying bargain-basement-priced goods made in China even if in the process they were shopping themselves right out of their own jobs.

As long as you weren't one of those blue-collar workers whose job was "offshored" to a cheaper labor market, you benefited from the new globalized economy. Your service-sector income could now buy more iPods or flat-screen TVs than ever before. By making everything cheaper, globalization meant you had more money to spend. And most of us in the developed world were more than happy to spend it. But the global market is also an artifact of the era of cheap energy. Moving your factory to the other side of the world to take advantage of lower labor costs makes sense only when goods can be moved around the world cheaply. Oil prices averaged only about $14 per barrel in today's dollars during the first big explosion in world trade between 1960 and 1973. The years between 1987 and 2002 saw another quantum leap in world trade, spurred not only by a 30 percent drop in tariffs but also by still relatively cheap transport costs. In today's dollars, oil was about $25 per barrel back then.

When energy is cheap, times are generally good for the world economy. But when the price of energy becomes dear, the world economy gets ice cold. The rising trend in exports as a share of world GDP stopped dead in its tracks between the first OPEC shock in 1973 and the aftermath of the second OPEC shock in 1979 despite a 25 percent reduction in global tariffs. Transport costs, particularly on long-haul transoceanic trade, tripled and nullified the impact of all the cuts in global tariffs that were going on at the time.

You can liberalize trade all you like, but it won't make a difference if no one can afford to ship the things you want to sell. High transport costs not only were a huge brake on the growth of world trade between the oil shocks, they also diverted trade along increasingly regional lines. As the cost of transoceanic freight quintupled following the 1973 OPEC shock and into the early 1980s, the share of nonpetroleum US imports from overseas fell by a stunning six percentage points in a little over five years, while the share of imports from the Caribbean and Latin America rose by a comparable amount. In terms of trade shares, those shifts are huge, involving the rerouting of billions of dollars of goods.

Change the price of a barrel of oil, and you change trading partners—if you trade at all, that is.

What lessons do the oil-shock-ridden 1970s have for us today? In 2008, no sooner had we had a taste of triple-digit oil prices than world trade stopped dead in its tracks for the first time in over two and half decades—when an earlier oil shock achieved the same result. A recession has already made China and its gleaming new factories look far away to Western consumers. And a return to triple-digit oil prices in the next economic recovery will make it seem even farther away. Are we also going to see a return to regionalized trade, making America's huge trade deficit with China a distant memory of a past world of cheap oil?

ROUGH SEAS

Container ships are the pack horses of global trade. You've seen them—the colossal oceangoing ships with what look like gigantic Lego blocks stacked on the decks. Inside those boxes are the fruit of globalization: all the things that are made on one side of the world and sold on the other. These floating warehouses run on the scrapings found at the bottom of a barrel of oil, an increasingly valuable gunk known as "bunker fuel."

The massive trend toward containerization has sobering implications for the future of global commerce. If anything, the world trade system may be even more vulnerable to soaring oil prices today than it was back in the 1970s. Recent changes in shipping technology have made transoceanic freight costs much more sensitive to oil prices than they were in the time of the OPEC shocks, suggesting we will see an even greater impact on world trade than occurred back then.

Container ships can be unloaded much faster than bulk cargo ships, so they spend much more time at sea than in port. An average commercial transport ship now spends 85 percent of its time at sea compared to 55 percent only fifteen years ago. During those years, the share of cargo shipped by container has risen from 35 to 75 percent.

Greater speed has also made shipping costs more sensitive to fuel prices. The shift to container ships has increased the importance of ship speed since they spend much more time at sea than at port. Over the past two decades, container ships have been built to go faster than bulk cargo ships, and since container ships have steadily replaced bulk ships, the world's fleet speed picked up.

But greater speed requires greater energy. In global shipping, the increase in ship speed over the last fifteen years has doubled fuel consumption per unit of freight—another head fake from the efficiency paradox.

That means that fuel costs have steadily risen as a share of total freight costs. As oil climbed to $100 per barrel, fuel costs became almost half of the total cost of shipping something by sea. The implication for the future is clear. There is a direct link between transport costs and every increase in world oil prices. Higher oil costs translate directly into higher shipping costs, often through what is called the Bunker Adjustment Factor, which adds the increases in oil prices directly to the shipping bill. The increase in oil prices from $30 per barrel to $100 per barrel raised the average daily fuel bill of a cargo ship from $9,500 to $32,000.

Chinese exporters to America have faced particularly steep increases. The transpacific bunker charge, a benchmark fuel surcharge on freight from China, rose from $455 to over $1,100 on a standard 40-foot container between January 2007 and 2008. As the price of oil goes higher, China becomes a costlier and costlier place to get your stuff from.

And those costs will in turn be absorbed in the prices of whatever you are buying at Wal-Mart or Tesco or whatever the high-volume, low-price, big-box purveyor of foreign-made goods is where you live. Until Wal-Mart finds a cheaper place to source those very goods, that is. In a world of triple-digit oil prices, a cheaper supply source will be a closer supply source.

All of a sudden China has never looked farther away.

THE BUNKER TARIFF

If rising oil prices seem like a minor speed bump in the path of the eighteen-wheeler of globablization, consider that these are just the sort of costs you would add to global trade if you were setting out to deliberately stall it.

Transport costs act just like a tariff, which is simply a tax on imports. Both are impediments to global trade. The higher the tariff, the less competitive those imports are in the market, since the whole point of the tariff is to make imports more expensive than domestically produced goods. It is not very difficult to calculate the tariff equivalent rate for a given increase in transport costs—just gauge how those costs affect final retail prices.

In other words, if transport costs double because of soaring fuel prices, we can calculate how much those costs raise the final selling price of the item being shipped and then express that amount as if it were a tariff.

So just how high would those tariffs be? We got a glimpse of that between 2004 and 2008, when world oil prices shot up from $30 to almost $150 per barrel. Transoceanic shipping costs tripled over the period. Express that in tariff terms, and the increase offsets all the trade liberalization of the last three decades. That's one hell of an increase.

Back in 2000, when oil prices were $20 per barrel, transport costs associated with shipping cargoes across the Pacific were the equivalent of an average 3 percent US tariff rate. At $100 per barrel, transport costs are equivalent to an average tariff of 8 percent. At $150 per barrel—in the price range seen in the summer of 2008—the tariff equivalent was 13 percent, which basically took us back to the average tariff rates of the 1970s. Back then, the tariff wall was so high that most major markets in the world could be supplied only by domestic producers.

When oil is at $100 per barrel, transport costs outweigh the impact of tariffs for all of America's trading partners, including even next-door neighbors and NAFTA partners Canada and Mexico.

Exactly how much trade soaring transport costs will divert from low-wage overseas suppliers like China depends ultimately on how important those costs are as a share of total costs. Goods that have a high value-to-weight ratio carry implicitly small transport costs, while goods with low value-to-weight ratio typically carry significant shipping costs. If you are shipping a 40-foot container full of diamonds, you might not worry about the freight. A container full of plastic toys destined for the dollar store is another story.

As it turns out, a surprisingly high percentage of what China exports to the United States falls into the freight-sensitive category. Furniture and steel—typical Chinese exports—fall broadly in the high-transport-cost category. In other words, freight costs represent a significant portion of the final selling prices of these goods. That makes the competitiveness of Chinese suppliers in faraway markets like North America and Europe highly vulnerable to the soaring cost of bunker fuel that powers the ships that connect them to their export markets.

In some industries, China had already lost its competitive edge. Take the steel industry, for example. We got a brief glimpse of the shift in global cost curves in the US steel industry just before the recession hit, and the subsequent collapse in auto sales sent steel sales plunging for both domestic and foreign producers. But the recession obscures some startling shifts in the US steel market that had become apparent the minute oil prices shot into triple-digit territory.

By 2007, competitiveness in steel had already shifted away from Chinese exports and back to North American producers. Soaring transport costs—first on importing iron ore to China from Australia or from halfway around the world in Brazil, and then on exporting finished steel overseas to North America—added as much as an additional $90 onto the cost of what was then $600 per ton of hot rolled steel. That more than offset the Chinese wage advantage on what, thanks to technological change, had become as little as an hour and a half of labor time for that ton of steel. For the first time in over a decade, made-in-America steel had become cheaper than Chinese imports in the US marketplace. Long before the recession blew up the US steel market, Chinese exports to the US fell 20 percent between July 2007 and March 2008— and US steel production was up over 10 percent over the same period. All of a sudden American steel producers were winning back their own home market. Who would have thought that triple-digit oil prices could breathe new life into America's rust belt?

But that is what triple-digit oil prices, and the transoceanic shipping costs that go with them, can do. And not just in the steel industry. From industrial pump parts to lawnmower batteries to home furniture, shipping costs are driving production back to North America from cheap labor markets in China and elsewhere. With transport and logistics costs soon to become even more important, padlocks will be taken off mothballed factories, and machinery that hasn't run for years will be getting a new greasing.

Take the furniture industry, for example. It was once a thriving and important employer throughout the American South.

But then the massive industrial parks around Dongguan, just north of Hong Kong, started churning out furniture for markets in the US and the rest of the world, much of it built from wood imported from North America. In the process, hundreds of furniture factories closed their doors in places like the Carolinas, throwing thousands of skilled woodworkers out of their jobs.

But furniture is bulky and takes up lots of space in those 40-foot containers. With every dollar rise in world oil prices, a rise that shows up immediately on the transpacific Bunker Adjustment Factor on your shipping bill, another furniture factory in Taylorsville, North Carolina, gets a reprieve from some faraway plant in China where workers earn a fraction of what American workers get paid. At triple-digit oil prices, the net Chinese cost advantage in furniture is so marginal that it will no longer be worth the twelve-week wait to deliver sofas from Dongguan.

That is not to say that companies are going to give up on the pursuit of cheap labor just because they can no longer rely on cheap transport costs. But instead of finding cheap labor halfway around the world, the key will be to find the cheapest labor force within reasonable shipping distance of your market. Not just cheap—cheap and close by.

Look for Mexico's maquiladora plants to get another chance at bat when it comes to servicing the North American market. Mexico's brief moment in the sun was cut short by competition from China's much cheaper labor costs. But in a world of triple-digit oil prices, Mexico's wages don't have to be as low as China's in order to compete in the North American marketplace. (And given the pending loss of its oil exports, the Mexican economy will certainly need that break.)

Compare, for example, how oil prices can especially change relative transport costs between China and Mexico. At $30 per barrel, American importers pay 90 percent more to ship goods from China than from Mexico. At $100-per-barrel oil, importers pay 150 percent more to ship from China. And the higher oil prices rise, the more expensive China gets. At $200 per barrel, it will cost three times as much to ship a container from China than one from Mexico.

These shipping charges are not trivial no matter how cheap Chinese labor is. If you own a widget factory in Chengdu, you are probably starting to worry about the extra shipping costs that come with rising oil prices. Looking at the transport costs that come with triple-digit oil prices, your competitive disadvantage would be the same as having to pay a 20 percent tariff on the widgets you export to the US while your Mexican competitor would be free to sell its widgets duty-free in the American market. Those are the kind of cost advantages that will move your widget factory from Chengdu to Mexico in a hurry.

And the same logic that will have companies relocating from China to Mexico will have others moving north from Mexico. Crown Battery Manufacturing recently did just that when it relocated jobs from its plant in Reynosa, Mexico, and repatriated them to its Ohio home base. If you are making a 29,000-pound battery to run underground mining equipment in southern Illinois, even a 2,000-mile trek from a plant in northern Mexico may now be too costly at triple-digit oil prices.

Of course, the very same transport costs work both ways, making America too far away to supply the Chinese market as well. But given the direction in which most goods cross the Pacific, the onus of adjustment will be much greater on China than on North America.

What already happened to Chinese imports in the US steel market before the recession is likely to happen in a whole range of freight-sensitive sectors in the economy once the recession is over and oil prices begin to rise again. As that process unfolds, those huge trade deficits with China will quickly melt away. And as that occurs, our hollowed-out industrial landscape will just as quickly fill in.

CHAPTER 6

THE OTHER PROBLEM
WITH FOSSIL FUELS

THE NINETEENTH CENTURY WAS NOT EXACTLY a golden age from a climate-change perspective. For millennia, human technology hadn't changed a great deal. Wheels, metallurgy, basic navigation, rudimentary medicine and a few other things that define civilization (with the notable exception of gunpowder) had been in the human repertoire since before recorded history began. And though many other things came and went between the time our ancestors started writing things down and the days of the powdered wig in Europe, those essential technologies didn't change all that much.

Then we figured out how to put fossil energy to work, and almost overnight, everything changed. Cities changed, the countryside changed, work changed, labor and capital changed. And, much more slowly, the climate began to change as well.

The factories that sprang up in Great Britain and then across Europe and North America first began releasing the carbon stored in coal and sending it up into the atmosphere. Though global emissions at the end of the nineteenth century were a fraction of what they are today, about half of the carbon dioxide released by the factories that Charles Dickens and William Blake immortalized for their filth is still up there, trapping the earth's heat.

But in one of the ironies in which history seems to specialize, the nineteenth century has furnished us with part of the solution to the problems it bequeathed to later generations: economic theory.

Not surprisingly, the moment in history when global trade really began to take off was also the spur to the scientific study of what makes economies work. Those were the days of the gentleman amateur scholar. Since people were really just figuring out, say, biology back then, you didn't have to go to university to become a biologist. If you were rich enough, you just set up a laboratory wherever you could find the space (perhaps at your country home) and got to work classifying frogs.

Money is generally more difficult to come by than frogs, so it was probably not a coincidence that the preeminent nineteenth-century economic theorist was a millionaire (in today's dollars). David Ricardo had already made his fortune on the London Stock Exchange when he read Adam Smith's *The Wealth of Nations* and became interested in the study of where wealth comes from.

Ricardo's idea of comparative advantage is very simple, and very persuasive. It goes something like this: if everybody does what they are best at, rather than what they are just *good* at, and does only that, everybody will be better off. You don't have to be the very best at anything. In fact, you don't even have to be better than your trading partner at anything. All that matters is that you produce what you are least worst at.

In 1817, when Ricardo wrote *On the Principles of Political Economy and Taxation*, both wine and cloth could be produced more cheaply in Portugal than in England. While it might seem that England should have got out of both the wine and the cloth businesses, Ricardo thought differently. Portugal had an enormous advantage in winemaking (and still does), but a relatively minor advantage in clothmaking. If the Portuguese had insisted on producing and exporting both commodities, they would have given up the opportunity to take full advantage of their winemaking superiority and of exporting their excess wine to England in exchange for cloth. Since they were better at making wine than cloth, the Portuguese would get more cloth in exchange for their wine than they would in exchange for their labor, so they would be better off. And the English would be better off, since they would have a market for their cloth. Plus they would get to drink Portuguese wine rather than English.

In any case, that is exactly how it worked out. The British textile industry dominated the world during the reign of Queen Victoria, and the British got to toast their success with port, so called because it was shipped to England from the city of Porto, where the names of several British exporting concerns still grace the old warehouses. Economists like to be proven right just as much as anyone else, and Ricardo really nailed that one.

But Great Britain did not put together an empire on which the sun never set by selling cloth to the Portuguese. What Britain did best—indeed better than anyone else at the time—was make good use of the energy previously locked away in seams of coal. Fossil energy has often been likened to an army of cheap slaves because it does so much work in return for the energy it takes to exploit it, and Britain mobilized those slaves through industrialization better than anyone else for a long time. Harnessed to an economic regime of liberalized trade, the use of fossil fuel allowed British industry to flex its comparative advantage and dominate the world.

Others eventually caught up, of course. France, Russia, Japan, the United States and especially Germany embarked on massive programs of industrialization (and coal consumption). By the outbreak of the First World War early in the twentieth century, the other countries had by and large erased the lead Britain had jumped out to. And they did so by emulating the United Kingdom's carbon emissions.

The result for the Western economies (and Japan) was unprecedented technological and economic growth that continues to this day, despite two world wars, a depression and a half century of military standoff with the Soviet Union. What went largely unnoticed was that burning all that coal (and, later, oil) was slowly heating up the planet. Not that no one noticed the environmental consequences of coal use—they were hard to miss, since Western cities were blackened with soot and choked with fumes. Particulate emissions from coal plants brought on fog in London so thick that it was often impossible to see across the street. But the steadily rising levels of carbon dioxide in the atmosphere didn't seem like much of a problem in comparison to the economic benefits they brought.

And so, as the world economy grew by leaps and bounds, so did the level of carbon dioxide. As we get richer, we slowly but inevitably make the planet less habitable. From a preindustrial level of 280 parts per million of atmospheric carbon dioxide, we have reached 390 parts per million today. Some climate scientists think that 450 ppm is the fatal threshold beyond which we are committed to catastrophe, but NASA's James Hanson, one of the world's foremost climate experts, thinks the figure is probably 350 ppm, a level we passed in the 1980s.

We seem to have become aware of the danger facing us at the very last moment. North America and Europe are becoming increasingly concerned about climate change and the role that man-made greenhouse gas emissions are playing in the process of cooking the planet. That concern has increasingly transcended party lines in the OECD countries and has even become something that both major US political parties can agree on. In Australia, environmental concerns swept aside conventional political wisdom as John Howard was tossed out of office even as his government presided over a period of impressive economic growth. His mistake? Ignoring the threat of climate change in one of the countries among the very first to feel its effects. Kevin Rudd stepped into the prime minister's office and promptly signed the Kyoto Accord. Fighting climate change has gone from fringe politics to the mainstream all across the developed world. It has become motherhood and apple pie.

While combating a global recession has temporarily put carbon issues on the back burner, those concerns will grow, not lessen, in the years ahead as we see more evidence of melting polar ice caps, summer heat waves in Europe and increasing drought around the world. There are a lot of great reasons to work toward a carbon-constrained civilization, and topping the list would have to be concern for the kind of world our children and grandchildren will inhabit. No one wants to bequeath to future generations a world marked by mass extinctions, famine and wars over food.

Unfortunately, that kind of world is looking more likely by the day. Total global emissions have risen by a cumulative 25 percent since the beginning of the decade. For all the carbon abatement in the developed world, there is certainly no evidence of any emission slowdown on the global stage. In fact, emissions are rising faster than the worst-case scenario examined by the Intergovernmental Panel on Climate Change (IPCC). Despite the fact that the world is quickly waking up to the problem, emissions in the first decade of the millennium are growing four times faster than they did in the 1990s.

If there is any good news, it is that only a small fraction of that increase came from the rich countries of the OECD—the countries that cast the seeds of this problem more than 200 years ago. In fact, emissions from the most advanced economies in the world have grown by only 5 percent between 2000 and 2007, one–tenth of the 50 percent increase seen in the developing world. And that was during a period when oil barely averaged $50 per barrel. At a triple-digit crude price, emissions, like oil consumption itself, will be falling further in those economies.

In OECD countries, where consumers pay the full price for a barrel of oil, triple-digit prices will do more to reduce greenhouse gas emissions than a hundred Kyoto Protocols. Fewer motorists on the road driving smaller vehicles, plus half-empty airports, are going to put a big crimp in North American oil consumption, and in the process put an equally big crimp in North American carbon emissions. The run-up in oil prices from $20 per barrel in 2000 to $70 per barrel in 2007 literally brought emission growth to a halt in most OECD countries, including the United States. Consider what a near doubling of those prices could do to carbon emissions in the future.

Even with the additional emissions from the return of some heavy industries like steel, the emission cuts that will come from the transport sectors in North America will be overwhelming. Hence, North America's carbon footprint will get fainter and fainter just as carbon footprints get heavier and heavier in China and the rest of the developing world. Just as oil consumption has peaked in the US, so too have US greenhouse gas emissions.

Ironically, the US may ultimately yet comply with the Kyoto commitments it never signed on to. Emission reductions will be mandated not by legislation or international treaties, but by soaring fuel prices at the pumps and comparable increases in the price of jet fuel. A smaller world is a less carbon-intensive world.

But the developed countries have not begun to decarbonize just by traveling less. They have done it by turning their backs on the fuel that set them up as industrial powers in the first place—coal. France, for example, gets about 75 percent of its electricity from nuclear generators, while Denmark and Germany are world leaders in renewables. The mix of nuclear and hydro-electric power in the Canadian provinces of Ontario and Quebec is particularly climate-friendly. The easiest way to get emissions under control is less exotic: just switching from coal to natural gas generators cuts carbon emissions roughly in half. And that is exactly what has happened across the OECD countries.

Although the United States sits atop rich resources of coal, state regulators are increasingly saying no to new coal-fired capacity. Even in Texas, a state not known as particularly green, public pressure led to the withdrawal in 2007 of applications to build as many as eight new coal plants that would have provided 6,000 megawatts of new power. And they are by no means alone. Between 2006 and 2009, applications for 83 new coal plants were either turned down or withdrawn in the United States.

In today's world of mounting concern about greenhouse gas emissions, the politics of building a new coal plant in North America are broadly akin to the politics of getting a new nuclear-powered facility approved just after the nuclear accident at Three Mile Island back in the late 1970s. No new coal-fired plant has been built in the UK since 1986. And three coal plants in the Canadian province of Ontario are about to be converted to burn biomass.

That the old carbon reprobates of the OECD are mending their ways can only be good news for the planet. Environmentalists are less likely to be pleased by what they see outside of North America and the rest of the OECD community. While the Kyoto Accord made global carbon management the business of those who have historically emitted the most, that's not where today's emissions come from—or, more importantly, where tomorrow's will come from.

INDUSTRIAL REVOLUTION 2.0

While we wrapped up our own industrial revolution a long time ago, another one is underway on the other side of the world. And the fact that greenhouse gas emissions in the developing countries have shot up 50 percent since 2000 shows that, this time around, the industrialists in China are proceeding with an enthusiasm the top-hat-wearing, cigar-chomping capitalists of nineteenth-century Britain and America could only envy.

The increase in emissions in the developing world dwarfs the decline in emissions in the developed world, just as the increase in oil consumption has dwarfed the cutbacks in oil consumption in North America and Europe. Since 2000, the developing world has contributed 90 percent of the total increase in global emissions. Yet these are the very countries that were exempted from the emission constraints of the Kyoto Accord.

So great has the recent rise in emissions growth in the developing world been that by 2005 those emissions surpassed the emissions from OECD countries. Moreover, the gap has widened since. Already, the non-OECD nations emit a massive 2.5 billion metric tonnes a year more than the OECD, equal to 20 percent of the latter's entire emissions. And that gap will get wider every year.

You have to burn carbon before you can emit greenhouse gases. As we noted in chapter 2, oil consumption in the developing world will soon surpass OECD consumption. When it does, the emission growth of developing countries will accelerate even further, leaving emissions from the developed world in their dust.

And that is just oil. The developing world has long surpassed the OECD when it comes to burning coal, the dirtiest means of generating energy from hydrocarbons. Coal is the most carbon-intensive fuel out there, emitting twice as much carbon per unit of energy as natural gas and about 20 percent more than oil. The higher oil prices climb, the more coal the developing world is going to burn, since many of those countries still burn oil to generate power. In other words, while rising oil prices may kill some demand, they will also shift much of the demand to coal. If we don't put a price on atmospheric emissions, coal-fired power is anywhere from two to three times cheaper than either natural gas or nuclear. It is not hard to see why developing countries in desperate need of power so readily succumb to the temptation to burn more and more coal.

It is a difficult temptation to resist, particularly if you are not a rich country. Coal consumption is expected to double between 2005 and 2030—and China and India together will create almost 80 percent of that increase.

That is an increase on top of a huge increase. Since 2000, China's greenhouse gas emissions have more than doubled. In 2006, China passed the world's largest economy, the United States, to become the single largest emitter country in the world, and is now poised to move much farther ahead of the American economy in that regard. By 2007, China's emissions had already grown 10 percent more than emissions in America.

When it comes to emissions, there are two critical characteristics of an economy. The first is how much energy it takes on average to produce a unit of GDP. The less energy you use, the less hydrocarbon fuel you have to burn—and hence the less carbon you emit. An efficient economy is a clean economy.

The second factor is the carbon intensity of that energy, which measures the carbon emission trail per unit of energy consumption. If you are using hydroelectric power like parts of Canada or New Zealand, you don't emit any. Burning coal, however, is another matter.

The goal for all of us is to use less energy, and to use cleaner energy. On both criteria, the Chinese economy, and most if not all of the developing world, is seriously disadvantaged.

Energy use per unit of GDP in the Chinese economy is four times that of the US economy. That is, it takes *four times* as much energy to make a dollar of GDP there as it does in the US. Moreover, China emits a third more carbon per unit of energy than the US, and almost double that of the European Union. Combine the energy intensity of the Chinese economy with its poor carbon efficiency and then throw in double-digit rates of economic growth, and you have a powerful cocktail for exploding emissions growth.

No country burns coal these days like China—the country relies on coal for almost two-thirds of its total energy needs and for approximately 80 percent of its electrical power. By contrast, coal provides about 50 percent of the electricity in the United States and less than that in most other OECD countries. The UK relies on coal about half as much as China for power generation. In Canada it is less than half of that—about 12 percent. The only OECD country that is close to China in coal use is Australia, which gets about 75 percent of its power from coal. But then, Australia is the world's largest exporter of coal, so it's only natural that it might want to burn some of that coal itself.

Not only does China rely on coal for almost all of its power needs, but those needs are growing exponentially. Already, there are more coal plants in China than there are in the United States, the United Kingdom and Japan combined. While environmental concerns have made it increasingly difficult, if not outright impossible, to get new coal-fired capacity licensed in North America or Europe these days, China is adding new generating plants at breakneck speed to try to keep pace with soaring power demand in its booming economy.

Between now and 2012, plans call for over 500 new coal-fired electrical generating plants in China. The emissions from those plants alone will effectively nullify all the emission cuts required of OECD economies under the Kyoto Accord, such is the enormity of the emission cloud that hangs over the Middle Kingdom these days. (Carbon dioxide is only one form of coal-derived pollution that plagues the country: the smog that enveloped the 2008 Beijing Olympics gives barely a hint of the public health crisis in China—or in Japan, which is downwind from the smokestacks of its revitalized rival on the mainland.)

The new coal plants are expected to account for about 80 percent of the increase in the Chinese economy's greenhouse gas emissions over the next half decade. And it's not just China, of course. Coal consumption is expected to nearly double in Taiwan, Vietnam, Indonesia and Malaysia between 2005 and 2030. India's coal-fired generating capacity will more than double over the same span.

There are no national boundaries or borders up in the atmosphere. While every country has sovereignty on the ground, none has sovereignty in the biosphere. China's emissions are the world's emissions. Their carbon dioxide is shared equally with countries that have imposed significant costs on their own economies in order to reduce atmospheric emissions.

The explosive increase from China and the rest of the developing world is a lethal challenge to efforts in North America and the rest of the OECD to restrict and ultimately reduce carbon emissions. As OECD countries move to impose greater penalties on their own emitters, their tolerance for the profligate carbon practices of their trading partners will diminish dramatically.

No country is going to sacrifice the well-being of its own economy for the sake of some environmental goal, no matter how urgent, unless it sees other countries making the same sacrifice. Both the US and the UK are promising 80 percent emissions cuts by 2050. But that could be economic suicide if the rest of the world isn't ready to cut emissions too.

The challenge is, of course, to induce change in the carbon practices of the developing world. The problem is that the developing world isn't particularly interested in changing those practices any time soon. There is not the same outpouring of public concern over global warming in countries such as China and India as there seems to be in places like North America and Europe. Yet people in China and India have every reason to be as concerned about the impact of climate change as people in OECD countries—perhaps more so, since most climate-change models predict much greater environmental stresses, particularly drought, in their part of the world if carbon emissions continue to increase. Both China and India are already facing devastating water shortages caused by shifting weather patterns. Climate projections suggest that the Himalayan glaciers could be gone by 2035, and along with them the meltwater that feeds Asia's rivers. Two billion people depend on the Ganges, Indus, Brahmaputra, Salween, Mekong, Yangtze and Huang He rivers for year-round water supply. The world's biggest emitter has a lot at stake in the fight against climate change.

Still, despite the fact that energy demand in the developing world is increasingly at a multiple of the rate in the developed world, the average person in China consumes one-tenth of the energy consumed by the average person in North America. So it's perfectly understandable that China would tell the OECD countries to come back and talk to it about carbon management in fifty or a hundred years when its citizens have more comparable levels of energy consumption per capita.

The only problem is, global warming is not going to stop in its tracks and wait for China and India to catch up.

THE CARBON TARIFF

Efforts in the developed world to restrict and replace coal-fired capacity seem downright quixotic when juxtaposed against China's (and other developing countries') coal-expansion plans. Whatever reduction in greenhouse gas emissions is achieved in the world's developed economies from switching away from coal will simply be overwhelmed by the increase in emissions from new coal-fired plants in China and the rest of the developing world.

Saving the world is a noble motive for going green. But there is another compelling reason to want a carbon-abatement regime in place as soon as possible. It is called good old-fashioned naked economic self-interest. If we can't agree to save the world for someone else's benefit, we might as well do it for our own.

That's admittedly not the way Washington has seen carbon policy in the past. Far from it. The recent Bush administration always dragged its feet on putting a price on carbon emissions, fearing that the American economy would suffer too much from the resulting increase in energy prices. But the fact of the matter is that the economy has *already* suffered from the high cost of energy. Just look what is going on in Detroit these days.

Having already paid the economic costs from triple-digit oil prices, why not reap some benefits from your consequent reduction in oil consumption and hence carbon emissions?

Putting a price on carbon emissions when your emissions are falling and your competitors' emissions are soaring is compelling economics. And what's even better, it's economics that can be wrapped in a very green label.

What is the point of shutting down a coal-fired plant at home if another is opening up on the other side of the same planet? The answer to that question takes us back to David Ricardo's theory of comparative advantage. Countries should do what they are best at. Just as everyone was better off when Portugal, rather than England, focused on turning grapes and sunlight into wine, the whole planet will be in better shape when the countries that are most efficient in burning carbon get to burn the most. That is where their comparative advantage lies in a world where emitting carbon carries an economic cost.

What the Kyoto Accord failed to recognize is that in a world where greenhouse gas emissions are unevenly controlled, the right to emit suddenly becomes a source of huge comparative economic advantage. Manufacturing will quickly migrate from that part of the world where emissions are controlled (and hence cost money) to that part of the world where emissions are uncontrolled (and cost nothing). In exactly the same way that manufacturing jobs flee high wages and high taxes, they also flee expensive carbon.

In other words, Kyoto turns Ricardo's theory of comparative advantage on its head. Emissions won't migrate to those who can most efficiently manage them but to those who are simply allowed to emit. And they just happen to be the least efficient emitters. It is comparative advantage by decree and one that flies in the face of where true economic comparative advantage really lies. That's why US opposition to the Kyoto Accord wasn't so much about environmental impact as it was about economic impact. Most Americans didn't think Kyoto provided them with a level playing field, and they were right. Their overseas economic competitors wouldn't have had to play by the same carbon rules that the accord would have imposed on American firms. In 1990, the benchmark year for the treaty, 70 percent of world greenhouse gas emissions came from inside the OECD. In 2007, that number was 50 percent.

Numbers like that should tell us that the world has changed and so must Kyoto-type accords if they want to do more than move jobs from advanced to developing countries. It is not a question of moral posturing or assigning blame. North Americans and Australians are responsible for about 20 metric tonnes each per capita, while Britons weigh in at about 10 tonnes each (though the British emit more as a result of air travel than anyone else). The moral high ground is just not available to the world's wealthiest countries.

But as David Ricardo showed nearly two hundred years ago, the developed world does not have to be absolutely cleaner than their competitors to reclaim comparative advantage. They just have to exploit their lead in carbon management—in just the same way that the British exploited their lead in coal use. And the way to do that is by charging a tariff.

It is surely one of the great ironies of the smaller world on the horizon that someone like Ricardo, whose name is often invoked in support of free trade, should furnish us with the idea of charging tariffs on imports, but the world has changed since the days when the items that dominated trade were cloth and wine. Tariffs distort comparative advantage by protecting domestic industries—but a country's energy mix distorts comparative advantage in a similar way. Dirty energy is cheap energy, and therefore a subsidy to the industries that use it. So the carbon tariff is really not a tariff at all.

It is a countervailing tariff because it levels the playing field rather than tipping it in favor of the home team.

The most direct strategy for halting the seemingly endless growth in global carbon emissions, is not another round of Kyoto talks calling for voluntary cuts. What we need to do is to impose a carbon cost on emitters at home, then impose the same standards on imports.

As simple as that. It may be expensive to get cleaner, but it is going to be a lot more expensive for our competition. Save the world and beat your trade rivals while doing it—can there be a more satisfying win–win than that?

There is already growing talk in Europe of going this route. Since 2004, Western Europe has imposed economic costs on its own emitters. Today, the right to emit a metric tonne of carbon emissions costs about €13, or a little over $20, on the European Climate Exchange. Those prices are likely to rise as mandated emission cuts in Euroland become increasingly more severe over time. (Emission credits were only recently trading as high as €24, or $40 per tonne prior to the 2008 recession.) When a country like France gets 90 percent of its electricity from low-carbon nuclear generators and hydro, it only makes sense for it to work toward a global economy that favors the countries with comparative advantage in carbon management.

As Europe raises its own bar on emissions, there are calls for a carbon tariff on imports from countries like China that don't play by the same carbon rules.

If European manufacturers have to pay to do the right thing, they shouldn't have to pay a second time by giving up competitiveness against trade rivals that underprice them by doing the wrong thing. And what is true in Europe is true in the rest of the developed world, including North America.

In effect, a carbon tariff could affect everything from making steel to manufacturing knock-off Rolexes. The greater the emissions that have gone into producing whatever good is being exported, the higher the tariff that will be charged when that good enters North American or European markets.

A carbon tariff, and the resulting restriction in market access it would bring, may in the end be the only way to get the rest of the world to manage emissions. If China emits greenhouse gases for its own domestic consumption, there is little we can do about it. But if China powers its export industries with carbon-spewing coal-fired generating plants, we can insist that Chinese exporters pay a tariff on those grounds.

As it turns out, China's export industries are a major and growing source of the country's world-leading emissions. Recent estimates suggest that a third of the country's total greenhouse gas emissions come from its export sector. That makes China's export sector alone the second-largest emitter in the world, next only to the total emissions from the US economy.

Meanwhile, if you live in the United States, you will soon be paying two or three times more for your electricity because your state regulator has forced the utility to burn natural gas instead of coal in order to save the environment from ever greater greenhouse gas emissions. What that means at an international level is that you will be subsidizing another country's trade advantage by going greener. That is as bad a deal for your own economy as it is for the global environment.

For that reason, trade policy analysts are going to warm up to the idea of a carbon tariff. But others, including the Chinese, will ask whether that is exactly fair. Didn't we export our own emissions to China when we sent factories overseas and then later imported the very goods that those factories used to produce at home? Can we now penalize China for the emissions that we exported just so we could use their cheap labor to make the things we consume cheaper?

Fair questions. But it does matter where the factories are located, because not all factories are created equal when it comes to carbon emissions. While a factory in China and a factory in the United States emit into the same biosphere, fundamental differences in their energy intensity and in the carbon intensity of the energy they use give rise to very different emission trails.

That's where economics can become very green.

Because the Chinese economy is so carbon intensive, it is about the least optimal place for the world's carbon-intensive industries to call home if carbon emissions carry an economic cost. If, for example, China had the same energy intensity as the US economy and its energy had the same carbon intensity, its emission growth since the beginning of the decade would have been one-fifth the 120 percent increase recorded to date. That in turn would have saved the atmosphere a staggering 2.7 billion metric tonnes in greenhouse gas emissions.

Now that concentrations of atmospheric carbon dioxide are approaching critical levels, we would love to have those billions of tonnes back. And as time goes by, we will pay more to get them back.

While it is true that the US exported much of its own emissions to China when it exported much of its heavy industry to cheap Chinese labor markets, that shift occurred when greenhouse gas emissions didn't cost anything. Once the market is allowed to put a price on those emissions, that shift may no longer make the same economic sense.

Simply replacing China's manufacturing exports to the States with American production would make a significant dent in global emissions considering that US industry can produce the same manufactured goods for roughly *half* the carbon emissions. And shifting production to Europe would be even cleaner.

Environmentalists don't get many chances to cut emissions in half with the stroke of a pen. But that is what would eventually happen under a carbon-tariff regime. And in the process, it would bring a lot of long-lost jobs back home.

Once again, we can feel the weight of global cost curves shifting, just as they moved in response to exploding transport costs falling out from soaring oil prices. In a world economy facing emission constraints, you want to locate emission-intensive industries not in the cheapest labor markets, but in the countries that have the most carbon-efficient technology, so that the world emits the least amount of greenhouse gases possible for a given level of economic activity. That's what global carbon-emission management ultimately comes down to: Ricardo's theory of comparative advantage.

The only reason that economies like China's can attract emission-intensive industries to their shores is that no one is forcing those countries to pay for the carbon emissions they belch into the atmosphere every day. In the parlance of economics, this is called a "classic market failure," arising from the fact that the market fails to recognize carbon emissions as an economic cost. The remedy is simple. Put an economic price on emissions, and the market will determine where those emissions will come from.

Once emissions carry a price, they work just like transport costs. The higher the costs of shipping goods from China to North America, the less important China's wage advantage becomes for whatever is being shipped. Similarly, the higher the price we put on carbon emissions, the less important the wage gap becomes in determining which side of the Pacific most emission-intensive industries will call home.

Given the size of the trade deficit with China, there should be no lack of motive for a higher carbon standard in the United States and Canada. And that's before even considering the positive fiscal impacts. At $45 per ton of emissions, the US Treasury is collecting a cool $55 billion, enough to finance all manner of green iniatives. All that is required is that the US apply the same carbon standard to its own domestic industries. After all, you can't call carbon emissions an unfair trade subsidy until you collect on those emissions from your own producers.

There are basically two ways you can put a price on carbon emissions in your economy. The first is through a carbon tax. This works just like a sales tax. You can apply it on just about anything whose production involves burning carbon and hence emitting greenhouse gases into the atmosphere. That means you can apply it to gasoline, or to coal-fired electric power, or even to plastics or fertilizer, since, like so many things today, both are made through burning oil or natural gas.

The other way, which featured among President Barack Obama's campaign promises, is through what is commonly referred to as a "cap-and-trade system." That's what Europe uses today, but the idea was actually pioneered in the US several decades ago in the regulation of emissions from power utilities.

In cap-and-trade, the government sets an overall environmental target by imposing a limit on the total amount of emissions that can be released from the power industry. You don't set up the rules and incentives and hope they work—you determine just how much should be allowed to be emitted, then let the emitters figure out the most cost-effective way to hit the target. Those who don't figure it out end up having to buy emissions credits from those who do. Then the market sets a price on those emissions by allowing utility companies to bid for the right to emit, and rewards companies that cut their emissions fastest and deepest.

Cap-and-trade was originally introduced in the United States in response to the acid rain problem that was threatening the Great Lakes as well as a number of other North American lakes and waterways in the 1970s. The systems proved remarkably successful in bringing about significant absolute reductions in the levels of nitrogen oxide (NO_x) and sulfur dioxide (SO_2) that were emitted by utilities. Moreover, as the price of emissions rose, they encouraged emission-reducing technological change, such as bringing down the cost of scrubbers that utilities could put on their smokestacks. But whether an operator buys emission credits or installs emission-reducing technology, putting a price on carbon emissions will have an immediate impact on economic behavior. Every ounce of carbon that goes up a smokestack will flow through to the utilities' bottom line, making shareholders very green in the process.

Whichever path to carbon pricing the United States takes, one of these days legislators are going to realize that the US economy actually has a comparative advantage in carbon emissions and should attempt to galvanize that advantage by putting an economic price on them. The higher the price of carbon emissions, the more American-made goods will replace overseas imports that are based on cheap labor and dirty carbon practices. And that's equally true for Western Europe, Japan, Canada and other advanced economies that employ much cleaner energy practices.

The carbon tariff won't stop China from burning massive amounts of coal. Two-thirds of China's coal consumption is for delivering power to its domestic market. But a carbon tariff will ensure that China's exports receive no commercial benefit from emitting greenhouse gases on world export markets. And that together with the impact of soaring oil prices on transoceanic transport costs could soon change the face of the world economy.

Will the new Obama administration in Washington warm up to the idea of a carbon tariff? Admittedly, most ambitious policy initiatives have not happened at the federal level. Washington and its Environmental Protection Agency (EPA) have always been laggards, not leaders, in setting the pace of US environmental policy.

From banning leaded gasoline to getting rid of PCBs, it's always been the state legislatures, not the EPA, that have done the heavy lifting when it comes to implementing new and tougher environmental standards and regulations. And it's certainly no different on the carbon front. Led by California, one of the largest energy markets in the world, most state legislatures in the US have already passed legislation to regulate their own carbon emissions. Most, if not all, have effectively banned the construction of new coal-fired generating capacity.

The story has been the same in Canada, where the province of British Columbia defied political wisdom by implementing a carbon tax while the federal government watched from the sidelines. Other provinces, including Ontario and Quebec, are also moving on the carbon front while the federal government waits to see what course a new administration in Washington will take.

Historically, when the US states lead, the federal government eventually follows when it comes to setting new environmental standards, so it is a good bet that the United States will soon be joining Western Europe in setting a price on its own carbon emissions. And when it does, it will have just raised the bar for all its trading partners, whether they want the bar raised or not.

How high emission prices will rise is still anyone's guess. Ultimately, that depends on how stringent the American environmental target is. The greater the reduction in emissions mandated by that target, the higher the price the market will put on carbon emissions. And the higher the price the American economy puts on its own emissions, the greater the counter-vailing tariff the US can apply on the dirty emissions embodied in most Chinese imports.

All of a sudden we have a level playing field. Chinese steel fac-tories can emit all they want, but once they ship their steel to the American or European market, they are going to have to pay, through a carbon tariff, the same carbon costs that domestic steel producers pay. Fair trade now means green trade.

At $45 per metric tonne, for example, the carbon tariff would be somewhere in the neighborhood of 17 percent. Throw in on top of that the extra freight costs that come with $100-per-barrel oil and the impact is equivalent to another 15 percent tariff. At $200 per barrel, the impact would be equivalent to a 25 percent tariff. In the smaller, greener world around the corner the total increase in costs is the same as if the US suddenly slapped an over 40 percent duty on Chinese goods.

That's a lot more than Senator Phil Gram was calling for in 2000 when he was advocating an all-out trade war with China. It's about fifteen times the average 3 percent tariff that Chinese imports are currently charged entering the American market. Yet those would be the tariff equivalent rates that Chinese exports would potentially face in a world of not only rising energy costs but also double-sided energy costs. It will soon not only cost a lot more money to buy oil, it's going to be expensive to burn it as well.

By putting a price on emissions, you will further alter global cost curves. The basic equation for international competitiveness is no longer just about the wage rate. It soon will be a much more complex economic equation about energy efficiency, carbon effi-ciency and transport costs. And tomorrow, the geography of where factories will be located in the world will be quite different from the economic geography we know today.

That doesn't mean, of course, that every industry that moved to China in the last twenty years is coming home. In many labor-intensive industries like footwear, textiles and clothing, there aren't enough carbon emissions and energy costs embodied in production to make a discernable difference to international competitiveness. And freight costs in those industries are too small for triple-digit oil prices to put the clamps on them. But in a whole host of other industries, ranging from chemicals to metal processing, putting a price on carbon emissions, just like soaring transport costs, will tip the competitive scale back in favor of North American industry.

STRANGE BEDFELLOWS

If you don't believe me, ask the unions.

The United Steelworkers of America and the Sierra Club may seem like strange bedfellows, but they are getting acquainted in a hurry these days. They've already forged a joint working group called the Blue Green Alliance. And while the notion of Archie Bunker getting into bed with Al Gore may still send shivers down the backs of both organized labor and the environmental movement, they had better get used to each other's embrace. A rapidly changing world will see their once diametrically opposing interests suddenly coalesce into a political coalition that will agitate for a cleaner environment while bringing long-lost jobs back to North American shores.

Just as soaring transport costs are a friend to North American labor, so too is carbon pricing. For once in organized labor's history, going green means saving jobs instead of sacrificing them.

That's still probably a hard sell at the union halls these days, but like a lot of things in our economy, big change is coming.

Most blue-collar workers don't have warm and fuzzy feelings about environmentalists. And for good reason. Usually, people with comfortable service-sector jobs are the ones who are willing to trade off blue-collar jobs in return for more stringent environmental protection. It's always easy to make green choices at someone else's expense.

But it is also easy to be green when it gives you a leg up on your competition. When unions go through the math, they will come to the realization that in a world where emissions cost money, the factories their members work in have a comparative global advantage. If your emissions are a third less than those of your competitors, you want the price of emissions to be as high as possible—because the bigger the cost of emissions, the greater your cost advantage from emitting less. That's not tree-hugging environmentalism, just plain old profit-maximizing economics.

Organized labor will soon need to advocate for more stringent standards for carbon emissions. In the process, blue-collar jobs become green-collar jobs. Not only does that change the way labor acts, it sets the stage for political alliances that would once have seemed absurd.

Save the world from disastrous human-induced climate change and in the process bring back home high-paying industrial jobs that most of us thought were gone forever: that's a powerful combo in any political constituency, and it's one that will soon agitate for a radical shift in US carbon policy. When American politicians finally figure that out, the US will quickly go from foot dragger to world leader on the carbon crusade.

After decades of watching ever-more-stringent environmental

regulation and standards send industries and jobs packing over-seas, the world is now about to change for labor. Just as soaring transoceanic shipping prices will turn global cost curves on their heads, carbon pricing can induce a 180-degree shift in labor's atti-tude toward environmentalism.

Armed with the knowledge that a third of China's carbon emis-sions comes from its export sector, US steelworkers have every reason now to be at the vanguard of those urging Congress for tougher action on carbon emissions.

American labor will soon learn that what has been lost through trade liberalization can be won back through environmental pro-tection. A carbon tariff, just like triple-digit oil prices, will soon be bringing a lot of long-lost jobs back home.

CHAPTER 7

JUST HOW BIG
IS CLEVELAND?

I NEVER KNEW THAT CLEVELAND WAS THAT BIG.

I'm sure it is big enough for the folks who live there. But big enough to blow up the entire world economy?

If you listen to the news lately, that is exactly what Cleveland is supposed to be doing. Boarded up, repossessed and now unsellable homes in depressed US property markets like Cleveland, all financed with easy-credit subprime mortgages, have hit world financial markets like some kind of super-toxic hydrogen bomb. Suddenly the whole world is bailing out.

Not only has the real estate meltdown blown up the two pillars of the US mortgage market—venerable Fannie Mae and her brother, Freddie Mac—but it has taken down some of the largest and most famous financial institutions on Wall Street. Century-old stalwarts of American capitalism like Bear Stearns and Lehman Brothers have been wiped out. Even Merrill Lynch and its symbolic bull had to be rescued and in the process swallowed by Bank of America, which is now suffering a bad case of indigestion.

How did Cleveland, once dubbed the mistake by the lake, get to pack so powerful a punch?

The pundits will tell you about how subprime mortgages got bundled up and sold as fancy financial instruments with exotic names like "collateralized debt obligations" (CDOs), and how these came to be enormously leveraged on bank balance sheets.

When the typical 24-month interest-free "teaser" period on your standard subprime mortgage came to an end, folks who couldn't afford a mortgage in the first place simply walked away from their homes. And when what little equity they had had been wiped out by the decline in housing prices, the process only accelerated. When people started mailing house keys instead of a check to their mortgage companies, some financial market derivative that somehow had a slice of that monthly mortgage payment defaulted. And as they did, the financial institutions that held those assets had to write down bigger and bigger losses until some actually went under. The more that wealth vanished from the banks' books, the less they had available to lend. And so the economy found itself caught in the jaws of a giant credit crunch.

Or so the story goes.

There's no question that the financial crisis that has arisen from defaulting subprime mortgages rocked Wall Street to the core. It certainly took a huge chunk out of my bonus. But when we are talking about the economy in recession, let's not confuse cause and effect.

It's pretty obvious how a poisoned market for financial derivatives blew up a lot of investment bankers, but it's far from obvious how defaulting mortgages in Cleveland caused much deeper recessions in Europe and Japan—where the bad news started long before the subprime problems lit up the US. To date, the recession is about twice as severe in the Japanese economy as in the US, and about 50 percent more severe in Germany. And it seems that the worse the news from Cleveland was, the stronger the greenback got. If the decimated Cleveland property market is really the center of the world economy's problems, why did everyone send their money there?

Those questions have me flipping the channel every time I hear the pundits explaining how the world economy's problems are about financial markets and subprime mortgages. Forget Lehman Brothers. There is something bigger going on.

OIL SHOCKS HAVE ALWAYS CAUSED RECESSION

Here's a clue. Soaring oil prices caused four of the last five global recessions.

The only one that wasn't caused by oil prices, the Asian meltdown of 1998, never even washed ashore in the major oil-consuming economies of the world like the United States and Western Europe. By contrast, two of the largest recessions in the postwar period came directly after the last two OPEC oil shocks. The second OPEC shock actually reverberated into two back-to-back recessions. A decade later, another oil shock, this time caused by Iraq's invasion of Kuwait and subsequent torching of many of its oil wells, also produced a fairly deep recession in 1991.

If you want to stoke the world economy, feed it a steady diet of cheap oil. And if you want to choke it, give it expensive oil. It will start sputtering in a hurry, and you won't need any help from Lehman Brothers.

Should it then really come as a big surprise that the world economy, or at least the economies of the world's largest oil-consuming countries, should have found themselves in a recession once again when oil prices spiked to as high as triple-digit levels over the first half of 2008?

If past OPEC shocks could pack that kind of punch, no one

should doubt the hitting power of the recent run-up in oil prices. The over 500 percent explosion in oil prices from 2002 to the summer of 2008 is almost *double* the increase that occurred during either OPEC oil shock. And that's measured in "real," or inflation-adjusted, dollars, so we are not comparing the much greater purchasing power of a 1974 dollar with a 2008 dollar. Even after adjusting for inflation, the increase in oil prices dwarfs all former spikes.

No wonder the US economy and the rest of the OECD are in recession. Yet, like in a Rodney Dangerfield movie, the biggest oil price increase of all time seems to get no respect when it comes to us recognizing what ails the world's economy today—or at least what ails the world's largest oil-consuming economies.

It is no coincidence that large oil-importing economies without subprime mortgage markets witnessed the same things happening in their economies that were happening in the United States. In fact, many of those economies are doing even worse than the American economy. While the US economy guzzles 19 million barrels of oil per day, it is also one of the world's largest producers, pumping out about 5 million barrels per day in its own right. Parts of the US economy, like Texas and Louisiana, are big winners from triple-digit oil prices. And Canada looked for a while as though it might be recession-proof, thanks to its ballooning energy revenues from the Alberta oil sands.

But Japan has no Texas or Alberta—it has to import almost all of its oil. Nor is there an upside to high oil prices for most of Europe, with the obvious exception of Russia and the odd North Sea producer. Oil-importing economies in Europe get whacked by triple-digit oil prices without any compensating pockets of new-found oil wealth in places like Fort McMurray to soften the blow. Their almost total reliance on imports underscores a vulnerability to triple-digit oil prices much greater even than that of the world's largest oil-consuming economy.

And even in the US, the bite of triple-digit oil prices, particularly on vehicle sales, has dragged the economy not only into a recession, but into what could be just as devastating a downturn as the ones caused by the first two OPEC oil shocks. Since this price shock is almost twice as large, don't be surprised if the current recession is even deeper and longer than ones caused by the cartel.

And if you look at the epicenter of this recession, you'll find it's not falling housing starts but plummeting vehicle sales.

Just like the 1973–74 and 1979–82 recessions. In fact, auto sales have regressed all the way back to the levels of those recessions. Those three-decade-low levels for vehicle sales aren't about subprime mortgages and defaulting CDOs. They're about gasoline prices that rose as high as $4 per gallon by the 2008 Memorial Day weekend, which suddenly left your average American paying more to fill up his or her SUV than to cover the weekly grocery bill.

The fact that oil prices can still trigger recessions may come as a surprise to many readers, who have heard repeatedly over the last three decades that the US and other developed economies are a lot less vulnerable to oil shocks than in the past. But just as the efficiency paradox has taught us that we shouldn't confuse efficiency gains with conservation, we shouldn't confuse lowering the oil requirement per unit of GDP with our economy's becoming any less vulnerable to soaring energy prices than it has been in the past.

It may take half as much oil to produce a dollar of GDP today than in the 1970s, but GDP is a whole lot bigger than it was three decades ago and, because of falling domestic production, the US economy is almost twice as dependent on imported oil as it was during the first OPEC oil shock.

Oil prices, not delinquent subprime mortgages, are what brought down the global economy.

CHEAP OIL BRINGS EVEN CHEAPER MONEY

It doesn't matter how great a bargain is: if you buy something with a credit card and make only your minimum payment each month, that new fridge or television or whatever it is you just bought is going to end up costing you a lot more than the number on the price tag. Just like distance costs money, *money* costs money. And just as cheap oil brought down the cost of distance, it made money a lot cheaper too.

In fact, for a while, people were actually giving money away. If you bought a car with zero percent financing, someone handed you some money. If you activated one of those credit cards that seemed not long ago to show up in the mail practically every day with ridiculously low introductory rates on transfers of your balance, a bank was giving away money. Ditto for mortgages around the world. In some cases, banks were actually forking over more to new homebuyers than the value of the home. Of course, lenders weren't doing any of that out of a new-found sense of generosity. They threw money around like that only because they figured they would get even more back. But they wouldn't have been throwing it around if it were hard to come by either.

For a while, there seemed to be more than enough money to go around. Just about anyone could buy the house of their dreams and drive a new car and spend far beyond what their salaries should have enabled them to, because there was always credit available to keep them afloat. Sure, that debt would have to be repaid eventually, but as long as interest rates were low, that mountain of debt didn't seem to be a problem.

But when inflation triggered by high oil prices started to increase the cost of that debt, servicing it started to look like a very real problem. The ratio of household debt to income, one of the key bellwethers economists look at in assessing household spending strength, has shot into the stratosphere. The increase in this measure reflects the extent to which credit has outpaced income. And the story is pretty well the same no matter where you go. American households saw their debt-to–after-tax income ratio rise from near 88 percent in 2000 to almost 125 percent by the middle of 2008. In Britain, the increase was almost as steep, jumping over 20 percentage points to roughly the same debt-to-income ratio American households are struggling with. In Australia, the ratio rose even more dramatically, soaring from just under 100 percent in 2000 to almost 160 percent by 2008, while in Canada, the ratio leapt from just under 100 percent to over 130 percent. In all four countries, not only did household debt grow measurably against income, it did so at an unprecedented pace. And the one thing that all these countries shared in common was a combination of record low interest rates and easy credit.

Consumer spending is good for the economy. That's why governments around the world are begging people to go out and buy fridges. But the thing about high oil prices is that while they raise the price of everything, they transfer those trillions of dollars of new money out of the oil-importing countries, where people spend their entire paychecks each month, to the oil-exporting countries, particularly OPEC countries, where savings rates are as high as 50 percent. Between 2005 and 2007, steadily rising oil prices transferred as much as a trillion dollars from non-saving OECD countries to high-saving OPEC economies. In other words, that money leaves and doesn't come back. It is the opposite of the fiscal stimulation programs we are now seeing all over the world. While the federal governments want to get money into the hands of people who are going to spend it, high oil prices hand it over to people who are going to sock it away.

Any central banker will tell you that your borrowing rate is a mirror image of the inflation rate. The higher your inflation rate, the higher the interest rate you will pay on borrowed money and the tighter the credit markets you will face. It's only logical— inflation means your money loses value. If inflation is running at, say, 5 percent, and I lend you $100, you would have to pay me back $105 just so that I can break even with the $5 decline in the value of my original $100 since I lent it to you. If it's at 10 percent, I'll need to get back $110. So the hotter inflation runs, the higher interest rates go. And the higher interest rates go, the fewer people will qualify for loans (since loans with higher interest rates are harder to pay back), and the less money there will be to go around.

Conversely, lower inflation means lower interest rates and easier credit conditions. Loans go to people who might not otherwise get them—people who finance a home through a subprime mortgage, for example. And not only do people who might normally not get a loan suddenly have credit at their disposal, they enjoy much lower interest rates than when inflation is high.

And why are inflation and interest rates so low? Globalization. We saw in the last chapter how cheap energy means cheap transportation, which in turn gives Western consumers access to cheap labor. And that means little inflation. It was cheap oil and the trade negotiators, not the central bankers like Allan Greenspan, that actually wrestled inflation to the ground.

Low interest rates come from low inflation, and low inflation comes from globalization. But globalization requires cheap transport costs, and hence cheap oil. As globalization expunged the world of inflation, central banks were free to bring down interest rates, which they quickly did. In that way, not only did falling inflation boost your purchasing power, it also fundamentally boosted your borrowing power. And in the end, the borrowing power that households got from easy credit may have been even more important than the purchasing power they got from low inflation when it came to fueling the economic boom that followed.

Under the auspices of cheap oil over the last three decades, inflation shrank from double-digit rates to as low as 1 percent in North America, and interest rates followed suit. In the 1980s, the federal funds rate—the key interest rate set by the Federal Reserve Board in the US—averaged almost 10 percent. In the 1990s that rate fell to an average of 5 percent, and this decade has seen it fall to 3 percent. By 2003, the federal funds rate had fallen to as low as 1 percent.

Record low interest rates and the easy credit conditions that they brought were essential for the development of the subprime mortgage market in the US. First, it was possible to offer unemployed homebuyers a subprime mortgage only in a world of easy and abundant credit. Otherwise their incomes—or more precisely the lack thereof—would never have met minimum mortgage eligibility requirements. When low-income workers were getting approved for subprime mortgages, regular folks with better-paying jobs were already getting preapproved credit cards in the mail with twice the spending limit of their existing cards. These types of extravagances exist only in a credit bubble, which in turn requires very low interest rates to finance.

Record low interest rates were not only essential to the supply of subprime mortgages, but also in generating the demand for such mortgages in financial markets. The attraction from exotic derivative instruments like CDOs, funded by subprime mortgage payments, were created by the record low interest rates that government securities like Treasury bills were paying.

The low inflation dividend from globalization and cheap oil was great if you were a borrower, since it meant you could buy your house with a minimal down payment. And they were great for investment banks that could leverage these mortgage-based derivative products, like those CDOs, that were funded from the monthly subprime mortgage payments. But it was anything but great for savers, who were now getting virtually no interest at all on their lifetime savings. Wall Street's leverage was the flip side of paltry interest rates on savings accounts throughout Middle America. Treasury bill yields had fallen to as low as 2 percent, and even a 10-year Treasury bond was paying only a couple of percentage points higher.

The desperate search for yield threw many a pension plan into the arms of exotic new financial derivatives that offered to pay 6 to 7 percent. And what's more, the rating agencies who were supposed to advise on risk were giving most of them an AAA rating, reserved for only the safest and most creditworthy investments, such as Treasury bonds.

In short, neither the supply of subprime mortgages nor the demand for high-yielding but risky mortgage-backed derivatives would have materialized in a world of normal borrowing costs and normal saving rates. But start giving away free money and before long, you are going to have a bubble economy, in either the housing market or for that matter the auto market, where about two-thirds of the vehicles are financed.

Take away the cheap energy, though, and the bubble bursts.

The credit bubble needed low inflation, and low inflation needed cheap energy. The problem was that oil prices started moving up so rapidly that inflation, and hence interest rates, would soon be quickly on the rise as well. That spelled tightening credit conditions at a time when many assets, particularly subprime mortgages and the fancy Wall Street derivative investments that they funded, were leveraged to the hilt.

From January 2004 to January 2006, the rise in oil prices from $35 to $68 per barrel drove energy inflation, as measured in the US consumer price index, from less than 1 percent measured year over year to as high as 35 percent. Together with an associated increase in food prices (which we will explore in the next chapter), soaring energy costs drove the overall consumer price inflation rate from below 2 percent to almost 6 percent during the summer of 2008, reaching its highest mark since the 1991 oil shock.

You don't have to be a Nobel Prize–winning economist to figure out what happened to interest rates over that period. As soaring oil prices stoked inflation's flame, the federal funds rate began a relentless climb from a record low of 1 percent to over 5 percent by 2007. And rates stayed by and large at that level for another year until the economy rolled over into recession. But just as interest rates were starting to catch up with inflation, a mountain of subprime mortgages came due for refinancing. Not only was the interest-free teaser period about to end, but the interest rates that subprime mortgage holders would now have to start paying were almost double the rates when they first got the mortgages. The rest is history. The unraveling of the subprime mortgage credit bubble has resulted in the worst postwar property crash in America and the worst banking crisis since the Great Depression.

But if oil prices had stayed at $20 or $30 per barrel, where they were at the beginning of the decade, inflation would have never risen and neither would interest rates. And if we had stayed in the world of easy money and easy credit, folks in Cleveland would still be in their homes. They were not the cause of the problem. They were the symptom.

GET OUT OF JAIL ALMOST FREE

Back in 1990, I was standing outside the Imperial Palace in Tokyo with my Tokyo branch manager. He had brought me there not only for a bit of sightseeing, but to illustrate for me a piece of trivia about the Japanese real estate market. The palace we were looking at was worth more than the whole island of Manhattan. That is a lot for one house. And to be honest, it is not even the nicest palace in the world. I mean, it is nice and everything, no offense intended. But it is no Buckingham Palace, or even a Loire château, for that matter. Still, it is in Tokyo, and on that day in Tokyo just about any piece of land was worth a fortune.

Of course, a lot has happened since then. The Japanese real estate boom turned out to be a bubble, and Japan is still trying to recover from the pop it made when it burst. For all of its techno-logical wizardry and impressive wealth, Japan has been stagnating for years, much to the surprise of those people who thought not all that long ago that the Land of the Rising Sun was on the cusp of world economic domination. But a pile of bad loans rotting in the portfolios of Japanese banks stank up the economy there so badly that, while the rest of the world was booming, Japan suf-fered through what is now called its "lost decade."

No one has to tell Fed chairman Ben Bernanke about the mys-terious ailment of the Japanese economy. Bernanke is an expert on the banking and credit crisis that led Japan into a decade of economic stagnation and deflation. And, judging by how aggressively the Federal Reserve Board has fought this recession, Bernanke also seems to be an adherent of the idea that those who do not learn from history are doomed to repeat it.

No one in the world is eager to repeat the Japanese experiment with a zero-growth economy. But the global economy can't grow at 4 to 5 percent a year with triple-digit oil prices. In fact, the global economy may not even be able to grow at even 1 to 2 percent. Unless, of course, it has a very strong wind at its back.

Find a strong enough wind and even pigs can fly, at least while it's blowing. Not including massive bank bailouts, governments around the world have committed over $2 trillion to fighting the global recession. The EU has thrown €200 billion at the problem so far, Japan has kicked in 12 trillion yen and China has written a check for 4 trillion yuan. Britain has so far pumped about £20 billion into the economy, while Australia is in for $28 billion and New Zealand for $5 billion—and so on around the world, with the US leading the charge. Not only has the Fed thrown everything but the kitchen sink at fighting this recession, including taking some of the worst of Wall Street's troubled mortgage-backed assets onto its own balance sheet, but Washington is about to unleash the biggest fiscal cavalry charge of the entire postwar period. Even before President Obama could set foot in the White House, the outgoing Bush administration's various bailout packages had pushed the US budget deficit to near $1 trillion. And with the new Democratic administration's stimulus plan, that shortfall rose to at least $1.7 trillion and possibly even more.

That's a huge deficit even for the $14 trillion American economy. As a share of the economy, it's 12 percent, dwarfing any budgetary shortfall that America has had to finance over the entire postwar period. You have to go back to 1945 and the huge deficits racked up over the course of the Second World War to see anything as large. Relative to the size of the economy, the federal deficit will be over twice as big as it was when Washington had to finance the Korean or Vietnam war.

But the enemy this time is not fascism or communism—it is economic stagnation.

If you force-feed a $14 trillion economy with $1.7 trillion of deficit spending, GDP growth will respond. That's not economic theory, just simple arithmetic. All that additional money will get spent in the economy, whether it comes in the form of tax breaks or government checks or public-spending programs. If only the Japanese had figured this out, so the thinking goes, they wouldn't have lost a decade of growth.

The question now is, would you rather be Japan—or Argentina?

When Argentina tried to spend its way out of decline, it ended up triggering a spike in inflation that peaked at 20,000 percent in 1989–90. The Argentine peso was worthless; the economy was in tatters; employment and productivity were in freefall. It is a fair bet that the citizens of Argentina would have been only too pleased to trade places with the Japanese.

Now, no one is saying the United States is going to go down the same path as Argentina. But that does not mean that the lessons learned from Japan are not another head fake. How long can you go on spending money you don't really have? What about the trail of record deficits that lies in the wake of such massive fiscal stimulus? How many years of painful program-spending cuts and endless tax hikes will it take to whittle down such a massive deficit? Future taxpayers might wish we had just bitten the bullet and absorbed the shock of triple-digit oil prices instead of hopelessly mortgaging their future to save industries that were already doomed by an energy crisis. You can't go on borrowing from the future when you have debts to pay to the past.

In the long run, racking up these huge deficits is bound to make things worse. The road back to fiscal solvency from massive government deficits is a long and arduous one. But history has shown that there are always tempting shortcuts, and the longer the austerity road back to fiscal solvency, the more those shortcuts will be pursued.

Already, the recession and the collapse in financial markets have challenged our faith in free trade and in laissez-faire markets. Whereas "government regulation" used to be dirty words, they are now a beacon for new policy from Washington to Brussels. But at the same time, these once fiscally prudent governments are throwing caution to the wind and abandoning the holy grail they cherished in the era of cheap oil and strong economies: balanced budgets. Not all that long ago, governments would be pilloried for running a deficit. Now governments seem to be competing to see who can rack up the biggest shortfall. In Canada, where balanced budgets were recently the goal of political parties of all stripes, the minority Conservative government faced being brought down at the end of 2008 by a coalition of opposition parties for not committing to a big enough shortfall—when only months before, the same opposition members had attacked the Conservatives for risking running a deficit. Just as the hard-nosed capitalists in Detroit turn into big-time Keynesians when the going gets rough, so too do politicians around the world, when it seems like only yesterday they were campaigning for fiscal austerity. Even central banks are changing their tune. After decades of fighting inflation, they have turned the printing presses back on to bring their old archenemy back.

Every time Washington has run up huge deficits in the past, it's ended up monetizing them and triggering a huge bout of resulting inflation. Monetizing the deficit, as its phrase suggests, means that the Federal Reserve Board simply prints more money in exchange for a whole bunch of Treasury bills and bonds that Washington sells to finance the budget deficit. In normal times, those Treasury bonds are sold to the public or to foreign central banks, such as the People's Bank of China, which hold them as part of their foreign exchange reserves.

When the public buys the Treasury bonds, the broad supply of money in the economy has not changed. As a result of holding more Treasury bonds, the public now holds less of some other asset in their portfolio, like corporate bonds, or stocks, or even cash. But when the Federal Reserve Board or any central bank buys its own government's bonds by simply printing more money and depositing the funds in the government's account, the money supply has all of a sudden grown. And that increase in the money supply gets distributed through bailout checks to the big three auto producers in Detroit, or to prop up money-center banks in New York, or to subprime mortgage holders back in Cleveland.

While all those handouts seem to make it easier for everyone who spends them in the economy, they actually make things harder over time. Because while the money supply has grown, the economy's productive capacity hasn't. There is just more money chasing the goods and services in the economy, and hence the price of those goods and services goes up. And the more inflation there is, the better off the government that is borrowing your money.

Inflation is always good for the debtor, because the loan is being paid off in dollars that are worth less every year. Remember our example of the $100 loan. In effect, borrowers have to pay back less if the debt is diminished by inflation. Of course, the creditor—in this case, the Treasury bond holder who is financing

Washington's budget deficit—feels very differently about inflation: while the bond matures at face value, inflation could have
eroded much of its purchasing power. Sure, you get the interest
you were counting on, but that and the money you get on matu-
rity won't buy what you hoped it would.

That is precisely what happened to Treasury bond holders the
last time they lent Uncle Sam the money to finance huge budget
deficits. Bondholders who financed the World War II deficits saw
their bonds lose nearly 15 percent of their real value in the
ensuing inflation that saw the CPI rate peak at around 14 percent
in 1947. Those who financed the deficits from the Korean War
weren't as badly fleeced, but, nevertheless, the resulting rise in
inflation knocked 5 percent off their real return. And twenty years
later, investors were once again swindled out of the return from
financing the Vietnam War. They lost nearly a third.

Monetizing deficits is particularly attractive for a country like
the United States, whose greenback is still the reserve currency
of the world. The fact that other countries want to hold your
money allows you to sell them bonds that are denominated in
your own currency. That gives the borrower a huge advantage,
because the creditor is at the mercy of the borrower's currency's
exchange rate. The easiest way to stiff a foreign creditor is to
simply devalue your currency. And the higher the inflation rate
you run, the more your currency will fall.

For the People's Bank of China—China's central bank and the
single largest holder of Treasury bonds—the risk lies with the
future value of the greenback. While the Treasury bonds they
hold will always mature at 100 cents on a dollar, a dollar could
buy a whole lot fewer yuan by the time it does, particularly over
the lifetime of a long-term 10- to 20-year bond. After all, the US
dollar lost 40 percent of its value against the yen between 1971

and 1981. That probably didn't go over well with all the Japanese financial institutions that held Treasury bonds over that period. Just as the domestic bondholder loses his return to inflation, inflation robs the foreign bondholder of his return through the fall in the value of the US dollar against the currency of the lender.

That's the beauty of being a reserve currency. You get to devalue at other people's expense. Better to rip off some foreign central bank than your own taxpayers. And it's an option that has become far more important to monetizing Washington's deficits than at any time in the past.

Back at the time of the OPEC oil shocks, only about 10 percent of the Washington's debt was owned abroad. If Washington was going to cheat its creditors, it was by and large American taxpayers who were lending it the money. Today, half of the US federal debt is owned abroad, much of it by central banks like the People's Bank of China.

Finally, the Federal Reserve Board has yet another compelling reason to hit the printing presses hard. Last, but by no means least, reflation helps to raise the value of all those financial market securities that have poisoned bank balance sheets. And some of the most toxic assets that blew up the likes of Bear Stearns and Lehman Brothers are now on the Federal Reserve Board's own balance sheet as part of its monetary reserves. That was the price the Federal Reserve Board had to pay to keep Wall Street and probably the entire global financial system afloat.

Higher inflation is bound to rub off on asset values, just as a rising tide lifts all ships. And one of the asset classes that is likely to be lifted by rising inflation is housing prices. Stop housing prices from falling, and all of a sudden more people are sending mortgage checks than home keys to their lending institution, and the valuations of all those exotic financial market derivatives that are leveraged to those mortgage payments will be better.

That means banks don't have to keep writing off billions of dollars from their balance sheets and can start doing what they are supposed to do, which is to lend money. And it means those otherwise worthless mortgage-backed assets that the Fed has taken on its own balance sheets might even be worth something one day, if the central bank can reflate the economy enough. If so, the Fed might even be able to sell back those troubled assets to Wall Street and recoup the billions of dollars of taxpayers' money spent to take them off the hands of failing banks.

When one outcome serves so many beneficial purposes, it is likely to be hotly pursued. While politicians and financial markets fret about the threat of falling into Japanese-style deflation, history has shown unequivocally that inflation is a far more common dancing partner for the spending spree going on in Washington and other economic capitals around the world these days.

Without fail, monetizing large government deficits has triggered massive bouts of inflation in the United States. The huge deficits that followed the Second World War saw monthly inflation peak at almost 20 percent in 1947. When the printing presses were turned on to pay for the Korean War, inflation jumped from negative territory to over 9 percent in less than a year. And when the Federal Reserve Board once again greased the printing presses during the Vietnam War, inflation soon made a triumphant return to double-digit territory.

The financial crisis that has imperiled the world's banking system is about to push the policy pendulum at central banks around the world from eliminating inflation in our economy to encouraging its speedy return. Fearing a replay of the crippling deflation that imprisoned the Japanese economy throughout the 1990s, governments around the world have already started to crank up the printing presses. And once the presses get going, they will be hard to stop.

RECESSIONS ALWAYS BRING OIL PRICES DOWN—BUT FOR HOW LONG?

Oil is in many ways its own worst enemy. Every time oil prices set new record highs, they deep-six the world economy. As demand contracts, oil prices come crashing down to the ground, just around the time that new supply comes gushing out of the ground, turning scarcity into glut. And of course, with prices having tumbled from their recent triple-digit perch, many are making the same argument today. But they are barking up the wrong tree.

Life is not linear, and neither is the direction oil prices move. Depletion doesn't mean that oil prices will always go up even if there will be less oil over time to burn. Price is always a function of both supply and demand. And while global supply is at best growing at a crawl, oil demand can fall if a recession is deep and long enough, like the one in 1979–82, and the one we are in now. And when demand falls, or even when demand growth slows enough, prices will fall. That doesn't mean that we aren't running out of oil. It just means a temporary reprieve until our energy appetite recovers and prices go on to new highs.

Most of the time, the global economy is growing. In fact, for the better part of this decade, it's been growing at a near-record pace, led by explosive GDP gains in the economies of China, Russia and Brazil. And since every unit of global GDP requires energy, strong economic growth translates into strong energy demand. The bigger the global GDP, the more oil we consume, as long as we power our economies on oil.

But the world economy doesn't grow all the time. Economies have always been subject to the ebb and flow of the business cycle, and that won't change just because oil is becoming increasingly scarce over time.

If your GDP shrinks, so does your appetite for energy. Look what happened in Russia in the wake of the collapse of the Soviet Union: energy demand plummeted by roughly a third as the economy simply stopped working. The United States is in no danger of vanishing overnight the way the USSR did, but in recessions people buy less and companies manufacture less. Consumers, corporations and governments all cut back on expenditures ranging from travel to how high they set the thermostat on cold winter nights. It's not just that people drive less—though they do—it's that they do nearly *everything* less.

Considering all the different ways in which a recession can reduce oil consumption, it shouldn't come as a great surprise that oil prices have tumbled from triple-digit levels to about $40 per barrel in the face of a global economic recession. But before buying that SUV, now that gasoline prices have fallen back to much more affordable prices, consider how long the recession is likely to last and what is likely to happen to oil prices once the recession is over.

Your average recession usually lasts three quarters. This one looks like a doozy and will last longer, but even the longest one on record in the entire postwar period lasted only six quarters. But over the last forty years, the US economy has been in recession only 18 percent of the time, and even the global economy has been in recession for only 20 percent of the time. What happens to oil prices when the recession is over? We will be facing the same pressure on oil prices that we were facing before the recession began. In fact, we will probably be looking at even tighter supply conditions than would otherwise prevail. While the media talk about how the recession is destroying demand for oil, the real lasting casualty is likely to be supply.

It's a lot easier to bring back oil demand than to bring back oil supply. Consumers don't need big lag times to rev up their oil consumption. Just cut pump prices in half and drivers will start filling up their tanks like they always have, provided they haven't lost their jobs. But while the plunge in oil prices will resuscitate demand, it's going to end up killing the very growth in supply that will be needed to meet that demand rebound.

From Canadian oil sands to deepwater wells off the Brazilian coast, scheduled new production is being canceled left and right as oil companies rein in spending in the face of falling oil prices. The plunge has suddenly made some of the world's largest oil megaprojects uneconomic. There are much easier ways to lose money than spending several billions of dollars developing a new Canadian oil-sand operation when extraction costs are more than double world oil prices. Over 1 million barrels a day of production has already been canceled in Alberta—where only a few months earlier the plan was to quadruple production. World markets were counting on that oil, and falling oil prices mean it is going to have to stay in the sand.

That supply growth may not be missed as long as a recession is suppressing global energy demand. But unless this recession turns out to be a decade-long depression, demand is likely to pop back up like a jack-in-the-box as soon as the economy turns the corner. But when it does, there won't be a comparable pickup in supply. Oil inventories will quickly plunge and oil prices will soar toward triple-digit levels again.

The key issue here is to separate the cycle from the trend. This recession has suppressed oil demand and brought down oil prices, like other recessions in the past. And no doubt future recessions will also see oil prices decline. But motorists have already probably noted that prices never seem to go back to where they once were. Instead, with each new recovery, gasoline gets more expensive. While $40 per barrel oil seems cheap by comparison to the triple-digit prices we have seen recently, it wasn't very long ago that would have seemed like pretty expensive fuel.

Oil is cheap in a recession only because you can't afford to drive. Maybe you've lost your job and that means you can no longer afford to make the monthly payment on your car lease, so the repo man comes and takes it away. That's why gasoline gets cheaper in a recession. But when economic life returns to normal, we can make those car payments and start driving again. And then gasoline prices will quickly climb back to the levels we balked at when filling our tanks before the recession.

Over the course of successive business cycles, the price of oil will move higher and higher—though not in a linear sense, since there will continue to be cyclical bumps along the way. And oil prices themselves will cause many of those bumps, as they caused the 2008 recession. But over successive business cycles, the price of oil will continue to rise, surpassing even the $147-per-barrel price seen briefly in 2008. The peak in one cycle's oil prices will become the trough over the next cycle. Yesterday $20 was a recession trough; today it's $40 to $45 per barrel—and in tomorrow's recession, it will be $80 to $100 per barrel. Each new cycle will bring not only higher peak prices but higher troughs as well.

In part, the rise in oil prices over the course of successive business cycles reflects the steady rise in marginal costs over time to produce oil as we increasingly scrape the bottom of the barrel. Oil has to sell for more on the open market when it costs more to produce. It used to cost $10 to pump a barrel of oil from underneath the North Sea. Today, now that those wells are in decline, that same barrel is going to cost anywhere between $60 and $95 to produce. There is no more $10 oil out there. The deeper we scrape, the higher the price we need to cover the cost of extraction and processing.

While oil demand will continue to fluctuate over the course of the business cycle, the inescapable fact is that the world consumes a lot more oil than it used to. Just as we are scraping the last dregs of world oil supply, the global economy has become thirstier and thirstier for oil.

Consider, for example, the last time world oil demand actually fell, back in 1983 in the aftermath of the double-dip recession, which, like today's, was caused by a huge rise in oil prices. Back then, the world consumed 58 million barrels per day and China consumed just less than 2 million barrels per day.

Today the world economy consumes some 86 million barrels per day, of which China consumes 7 million barrels. While it is certainly conceivable that the recession can temporarily reduce global consumption, including China's, China is not going back to consuming under 2 million barrels per day any more than the world is going back to consuming 58 million barrels per day, no matter how severe this recession turns out to be. In fact, a key part of the huge economic stimulus package that the Chinese government recently unleashed to fight the economic downturn involves massive spending on new road construction. And that's not to accommodate more rickshaws and bicycles.

More or less the same story can be told of oil consumption for Brazil or India or any of the other new oil-consuming nations of the world. Oil consumption in India has more than tripled since 1983, from barely over three-quarters of a million barrels per day to over 3 million barrels per day. Ditto for Brazil, whose daily oil consumption has risen from just under a million barrels per day to almost 2.5 million. Sure, oil consumption could fall as the recession spreads across the globe to places like India, Brazil and even China, but even with their economies contracting, they are still going to be guzzling a whole lot more oil than they did in the past. No recession can turn their oil consumption back to 1983 levels.

While higher oil prices can still cause cyclical changes in demand, as they always have, those changes have become less and less important in the face of long-term structural growth in oil demand in many newly consuming oil economies. In other words, how much oil demand falls in a recession becomes less important than how much it is growing over time. It's a little like playing table tennis on a moving train. The ball may seem to be bouncing back and forth, but really it is going forward the whole time.

Unless you get off the train, of course. It would be a mistake to look at slowing demand and falling prices and come to the conclusion that our recent brush with triple-digit oil prices was nothing more than a bizarre spike. The recession doesn't change the fundamentals—though it may make them harder to decipher for a while. But a temporary dip in demand won't make things better, and it's certain to make things worse if it pushes us back to our old energy habits. Recession can slow demand, but it doesn't slow depletion,

In the early 1980s, world production lost barely 2 million barrels a day each year from depletion; today the world loses roughly 4 million barrels a day of production every year, and that rate is accelerating. Considering that we have to replace nearly 20 million barrels per day over the next five years just to offset depletion, the oil industry won't have to worry about any future oil gluts like those that followed the OPEC oil shocks. We may wake up one day to the news that the economy is back on track only to discover that there is less oil supply at our disposal than there was when demand started to fall.

And it won't be just the one-two punch of reviving demand and sagging supply that pushes prices up in a hurry. Once the genie of inflation is out of the bottle, it is going to take oil prices on a ride along with everything else. For one thing, there will be more money chasing fewer barrels in the world so the price will go up.

And the dollars chasing that oil are going to be worth less and less even as the oil gets more valuable. Remember the Argentine peso and its 20,000 percent inflation rate? If a barrel of crude had been denominated in pesos, oil would have gone up 20,000 percent in 1989–90. If the United States wants to reflate its way out of recession, it is going to pump up the price everybody in the world pays for oil, since everybody pays in US dollars. If the dollar is worth less, oil is going to be worth that much more.

A WORLD OF SLOWER GROWTH AND HIGHER INFLATION

If triple-digit oil prices are the true culprit behind the recent recession, what happens if oil prices recover to triple-digit levels or even close to them when the economy recovers? Does the economy slip right back into recession again?

Everything else being equal—or *ceteris paribus*, as they say in the economics textbooks—that's probably as good a forecast as any.

Every oil shock has produced a global recession, and the record price increase of the past few years may produce the biggest one of all. But recessions, no matter how severe, are finite events. Ultimately, we face a far more challenging economic verdict from oil. Any way you cut it, a return to triple-digit oil prices means a much slower-growing world economy than before. And not just for a couple of quarters of recession.

That's because virtually every dollar of world GDP requires energy to produce. Not all of that energy, of course, comes from oil, but far too much does for world GDP not to be affected by oil's growing scarcity. And there is nothing at the end of the day that we can do about depletion.

Big tax cuts and big spending increases can mitigate triple-digit oil's bite, but the deficits they inevitably produce ultimately lead to tax hikes and spending cuts that just make the suffering all the more painful down the road. Taking out a loan to pay your mortgage might defer your problems for a month or so, but in the end, it often makes your difficulties more acute. Borrowing from the future just turns today's problems into tomorrow's, and by the time tomorrow comes, they've become a lot bigger than if we had dealt with them today. Trillion-dollar-plus deficits, just like a near-zero percent federal funds rate, can mask the impact of high energy prices for a while, but ultimately they can't protect economies that still run on oil from the impact of higher energy prices and the toll that they take.

There is only one way to avoid a future of much slower economic growth in a world of depleting oil supply. And that's to lessen the economy's dependence on oil.

That solution doesn't rest with the US Treasury or the Federal Reserve Board or the chancellor of the exchequer or the minister of finance. None can produce oil, and they can't wean the economy off oil either. They might be able to make things a little less painful, and they can definitely make things more painful. But they are not going to solve the problems posed by the dwindling supply of cheap oil.

Scarce oil can cause recessions and inflation only if we insist on consuming as much as we always have when it was much more abundant and cheaper. The key to downsizing the role of oil in our economy is micro decisions made every day by households and consumers, not macro decisions made at the level of monetary or fiscal policy. And that is just as true everywhere else in the world as it is in the country that consumes the most oil, the United States.

The plunge in oil prices is the reprieve we need to start weaning our economies off oil before we get clobbered with even higher prices than we did in the first half 2008. It's not just that Japanese-style deflation stifles growth, or that Argentinean-style inflation makes it impossible. It's not just that recovery from recession will trigger a spike in oil prices that in turn could provoke another recession. What is at stake is nothing less than economic growth itself and hence our very standard of living.

What the record plunge in oil prices from $147 per barrel to $40 per barrel over the recession clearly shows is that oil consumption and economic growth go hand in hand. As long as every new unit of global GDP requires someone somewhere in the world to burn more oil, the ability to grow the global economy is constrained by the ability to grow oil supply. If we can no longer grow oil supply, we will no longer be able to grow the economy—unless, of course, we can change the basic equation that ties the size of our economy to how much oil we burn.

That's certainly not going to happen with economic stimulus packages that are giving life support to dying oil-sucking industries like autos and allocating billions of new spending to road infrastructure. If we are going to go big time into hock with record-size government deficits, let's at least spend the money on our future, not our past. Our future is public transit, not more freeways for gas-guzzling private vehicles. If the public purse is going to support the auto industry, those public dollars should be spent on developing vehicles that aren't dependent on oil. Let fiscal stimulus be directed at the true cause of the recession, which is not Cleveland housing prices but the impact of triple-digit oil prices.

There is little we can do to prevent oil prices from recovering to ever new highs in the future. But there is a lot we can do to make sure that when they do, they won't have the same devastating impact. We have to change the basic equation that ties our oil consumption to our economy.

If we don't, even this recession may look minor compared to the economic future that awaits us.

[CHAPTER 8]

GOING LOCAL

I LOVE COFFEE ABOUT AS MUCH AS THE WORLD LOVES OIL. As it turns out, oil and coffee have more in common than meets the eye. Both have spurred the development of a global economy.

It all started with a humble 15th-century goatherd. As the story goes, this observant pastoralist noticed that his goats were noticeably friskier after eating the berries from a particular bush. Surrendering to the curiosity that has led to many great culinary discoveries (and not a few accidental deaths), the goatherd popped a few of the berries into his own mouth, and felt the better for it.

History does not record whether he immediately made himself a double-shot latte, but he did report the discovery to his imam, who set about devising a method of drying the berries and brewing them into a hot drink capable of keeping him up all night. Whether or not this is strictly true, it is known that by 1470 the citizens of Mocha, in Yemen, were drinking something called *qahwah* in Arabic, and that by 1510 they were drinking it in Cairo and Mecca as well. In 1610, a British traveler in Palestine noted that the locals seemed to spend all their time in what he called

"Coffa-houses," drinking a mysterious concoction "as hot as they can suffer it; blacke as soot, and tasting not much unlike it."

Before the global economy ran on oil, it ran on coffee.

When I go into the Starbucks near my office in Toronto's financial district, I am following a time-honored tradition. Before Europeans developed the habit of drinking coffee in the morning, they drank beer for breakfast—with predictable results. A beery breakfast may fill your stomach, not to mention your waistline, but it will do little to clear your head. And the minds of 17th-century London required great clarity. It was a time of rapid innovation and even revolution in nearly every field: politics, philosophy, science and, perhaps above all, finance.

Both the great failures and enduring successes of the early days of capitalism played out in London's coffeehouses. The notorious South Sea Bubble—a fraudulent investment scheme that has provided fodder for generations of economics textbooks—unfolded largely in a coffeehouse called Garraway's. A coffee shop that opened in 1680 created a legacy that has lasted much longer. Called Lloyd's, it was a meeting place for ship captains, investors and merchants. It was the place to go to learn the latest maritime news, and it was not long before Lloyd's was selling this information in the form of a newsletter. Eventually, insurance underwriters started renting out space in the café. Ninety years after it opened its doors, the coffee shop had been transformed into Lloyd's of London, today probably the most famous insurance company in the world.

One of the world's first and most important bourses also got its start as a coffee shop. London traders began meeting over coffee at a place called Jonathan's in the 1690s. In 1762, the arrangement became a little more formal: traders had to pay a subscription to do business on the premises. In 1773, a group of traders splintered off to do business elsewhere, calling the new business New Jonathan's. But they soon renamed the shop The Stock Exchange—the ancestor of today's London Stock Exchange.

Science, commerce, democracy—coffee was percolating through many of the developments that define the modern world. And there is one it affected even more crucially: globalization.

For those European coffeehouses to be kept buzzing with caffeine, a lot of beans somehow had to make their way from the plantations near Mecca to the port of Jidah, from which point they would make their way by ship up the Gulf of Suez and on by camel to Alexandria, where French and Venetian traders would buy them at auction and have them sent back across the Mediterranean. By the end of the 17th century, European countries had begun to worry about their dependence on Middle Eastern coffee and set about looking for a solution. Sound familiar?

The Dutch and then the French managed to break the Arab monopoly and establish their own plantations wherever their empires provided the right climate. If you could grow sugar, you could grow coffee, and that is just what they did. The Dutch set up their plantations in Java, the French in Martinique, which provided a jumping-off point for the bean's colonization of Brazil, Venezuela and Colombia. Today Brazil is the world's leading coffee grower, while the Arab countries have long fallen from prominence.

Of course, the Middle East has found something to handsomely replace its lost dominance of the coffee market. The trouble is, when that too comes to its inevitable end, how are coffee beans from Brazil or Java going to get to my local Starbucks?

Coffee has always been an artifact of a global economy, and that is no less the case today. Climate change may now allow you to grow cabernet sauvignon grapes in England and southern Ontario, but it is going to take a whole lot more global warming before I can grow coffee beans in my backyard.

That is not to say that coffee is going to suddenly disappear from the smaller world that is just around the corner. But without cheap oil, getting my coffee beans from halfway around the world is definitely going to cost a lot more.

Will drinking your morning coffee soon become as obsolete as drinking your morning beer?

THE END OF THE BARISTA ECONOMY

Your grandchildren may never know what a barista is.

All over the developed world, well-educated and cheerful young men and women with exquisitely developed expertise in making Americanos and soy-milk cappuccinos smile back at us from behind their counters. They serve an important need in our societies. But while our requirement for coffee is not going to vanish overnight, the market for it is in for a massive correction in a world of triple-digit oil prices.

Today the service sector makes up about 70 percent of the North American economy, and accounts for roughly the same proportion in the rest of the developed world. But it hasn't always been that way. In 1960, it was only 50 percent of America's GDP. The same proportions prevailed in similar economies such as the UK's and Canada's. Back in those days, there were far fewer people making sandwiches or greeting people in stores—or consulting or designing things, for that matter. But more people were busy making and fixing things.

One of the effects of globalization has been the expansion of what might be called the "barista economy." As we started importing more and more goods from cheap-labor markets overseas, the service sector became bigger and bigger. It was the one area of the economy where you didn't have to worry about competition from cheap labor overseas.

American dentists didn't have to worry about competition from Indian dentists because their customers weren't going to fly from New York to New Delhi to save a couple of hundred bucks on a root canal, any more than a Londoner would fly to Bulgaria to save a few pounds on a haircut or a manicure. And the only place I would consider getting my morning coffee is a few steps from my office building.

The beans may be global, but the barista is always going to be local.

The spectacular expansion of the service sector during the postwar period finds its counterpart in an equally massive implosion of the goods sector in our economies. Nowhere is that more evident than in food production. Agriculture's share of the workforce today is a fifth of what it was half a century ago. While not as dramatic a decline, manufacturing's share has fallen from about 30 percent of the workforce to about 10 percent. Whereas in 1960 roughly half of America's workforce was employed to produce some kind of good, today's employment in the goods sector has shrunk to about a third of the labor force. Just as fewer and fewer of us work in factories, manufacturing contributes less and less to GDP.

But the service sector could never have reached its current size if our economies had had to remain self-sufficient in cars, food, steel, televisions, computers and all the other things we now import. We became service-based economies precisely because we can now afford to import so many of those goods that we previously had to produce for ourselves.

As we have seen, we have been able to afford all these foreign-made goods because cheap transportation costs gave us access to cheap labor markets. Basically, we have been able to afford the services we have been providing to each other, from legal counsel to pedicures, by spending the savings generated by sending our factories overseas.

Now that the cost of importing goods from faraway places is soaring, we are going to have to roll up our sleeves and start producing our own flat-screen TVs for a change. And those baristas are going to be needed back in the factories, where corporations will soon be sweeping the cobwebs off the machinery. Where old factories no longer stand, new ones will be built. Capital, like water, flows quickly from high to low. If producing steel in the US makes more sense than importing it from China, new steel plants will emerge in America and elsewhere in OECD economies as quickly as they once disappeared. The services many of us are going to be providing in a smaller world will involve actually making things.

That means the once expanding service sector will soon start shrinking. At a minimum, it will have to give up labor, if not capital as well, to a suddenly growing goods sector of the economy as soaring transport costs force us to become more self-sufficient in things we thought we would never make again. Hence, the service sector's share of employment, just like its share of GDP, may start to fall and in the process begin to reverse what had been previously thought to be an irreversible trend in our economy.

This transition will not be a lay-up. The infrastructure, the technology, the training, even the work culture will have to undergo a massive overhaul to be anywhere near ready for a local economy. Many people who have not seen a lunch box since middle school, or who haven't had a callous on their hands since the time they tried to build their own backyard fence, may soon become reacquainted with both.

All of a sudden, the globalizing forces of the last three decades will come to a screeching halt. While trade liberalization and technical change have flattened the world, the soaring prices for energy are going to make the world rounder again. And this smaller, rounder world is going to look more like the past than what we are accustomed to expect from the future.

EATING OUT

There is hardly a sector of the economy that will not be affected by these global shifts. But perhaps nothing will affect our cities and our lives more than the fate of something we take pretty much for granted: our food.

My son's home hockey rink sits on what used to be farmland when I was growing up. I remember that the back fence of my public-school playground used to border on an open field.

Forty years later, there are 25 miles of suburban sprawl between that schoolyard fence and the nearest field. Farms are as hazy a memory as the Toronto Maple Leafs' last Stanley Cup parade. A whole lot of satellite communities built on what once was prime agricultural land are attached to Toronto through a complex array of interconnecting freeways. Around the world, just as in Toronto, cheap energy costs have allowed cities to gobble up the land surrounding them and to spread out in any direction available.

A notable exception is London, where planners had the foresight back in 1935 to surround the city with a development-free Green Belt, thereby limiting sprawl and in the process ensuring that London real estate prices would be among the highest in the world. Toronto adopted a Green Belt policy in 2004, but only after the city had lost 1.7 million acres of farmland, or about 35 percent of its original endowment.

The goalie crease my son is defending tonight was probably a tiny piece of a farm thirty years ago, growing crops that were sold in supermarkets that my parents shopped in.

So, with all the agricultural land gone, why am I not hungry?

I'm not hungry because I no longer depend on local farms to provide my food. The supermarket I shop at has food trucked, shipped and even flown in from all over the world. The food we typically buy travels anywhere from 1,500 to 3,000 miles. Average farmgate-to-store distance had increased by about 25 percent between 1980 and 2001. And the vast majority of that food had to be refrigerated when it was shipped.

Take the lamb chops sitting in my local supermarket's meat freezer. They're not from Ontario. Any grazing land within sight of Toronto was paved over a long time ago. They're from New Zealand, a country on the other side of the world that long ago read David Ricardo and decided to specialize in raising sheep.

And those blueberries? While they are indigenous to Ontario, that's not where the blueberries in my supermarket come from. Those oversized and probably genetically modified berries come from California, over 2,000 miles away.

That's a long way to schlep blueberries, and it's an even longer way to schlep frozen New Zealand lamb—to a place that can raise its own sheep and where blueberries grow wild.

The share of imports in our food supply has doubled over the last 20 years, even for fresh produce like fruit and vegetables. As we have become more and more integrated into a world food system built around global suppliers, the ability of local economies to feed themselves has steadily diminished. With cheap transport costs allowing you to enjoy the produce of farmers around the world, it no longer made any economic sense to be self-sufficient in food. Somebody somewhere else could always grow food and get it to your dinner table cheaper than your local farm.

Just as in manufacturing, globalization has turned agriculture into big business serving big global export markets. The source of comparative advantage in global agriculture comes down to the same labor-cost factor that comparative advantage in manufacturing is based on. So it shouldn't come as a big surprise that the same countries that have increasingly supplied the world with cheap mass-produced manufactured products should increasingly be providing the world with food.

While electronics and steel may garner all the press attention, food exports from China have managed until recently to stay under the radar. Nevertheless, the recent growth in Chinese food exports has been enormous. Everything from bok choy to apples to breaded chicken fingers is being sent around the world. Food exports to the US have soared from a modest $1 billion to $6 billion in 2008—a 500 percent increase. This brings a whole new meaning to having Chinese food delivered.

The same is true across the developing world. The global market for fruits and vegetables has grown dramatically during the cheap-oil era, doubling in size during the 1980s and by another 30 percent in the 1990s. As a result, food exports now make up about a fifth of the developing world's GDP.

Those food exports are even more dependent on cheap energy than China's steel exports. Steel doesn't have to be refrigerated when it is crossing the ocean. Much of the food China exports does. The energy and carbon trails from China's food exports are enormous. As in the case of steel, with every rise in oil prices, the cost advantage of farms in foreign countries becomes less and less important.

So forget about that avocado salad in the middle of the winter. If you live in Toronto like I do, the cost of flying in avocados from Mexico or southern California in February will soon make that salad prohibitively expensive. Air travel burns forty-four times as much energy as shipping does. All of a sudden, locally grown carrots and beets are coming back into style. The cost of smothering pancakes in Los Angeles with Ontario maple syrup may not make a whole lot of economic sense either.

In the process, the world just got smaller for both avocado growers in southern California and maple syrup producers in Ontario. The cost of transporting goods eliminates markets for both. And it won't just be avacados and maple syrup that will get the boot from distant consumers.

Downstairs in the mall of the office building where I work is a food court that could be the United Nations of greasy food. There are Thai, Indian, Chinese, Tex-Mex and Japanese food stalls. Virtually nothing served in the disposable packaging at any of these booths is grown anywhere remotely close to Toronto. Even the bottled water is flown in across the Atlantic Ocean from some spring in the Black Forest in southwest Germany.

Meanwhile, in the UK, where nearly two-thirds of the country's orchards have been chopped down in the past thirty years, 76 percent of the apples in the supermarket are from overseas, many from the United States, over 10,000 miles away. And though the lettuce in a British shop is more likely to come from relatively nearby Spain (about 1,000 miles away), it still takes 127 calories worth of aviation fuel for each calorie that salad yields in food energy.

The situation is the same in Australia, which is both distant from nearly everywhere else and geographically as spread out and car- and truck-dependent as North America. One Australian study found that a basket of twenty-five familiar supermarket goods had traveled over 43,500 miles from farm to table—nearly twice the distance around the earth. (A significant part of that— over 12,400 miles—came just from getting around within Australia.) Are these kinds of culinary choices sustainable in an economy of triple-digit oil prices?

In the new world of expensive oil and carbon tariffs, global cuisine, with its reliance on exotic imported ingredients, will soon be on the way out. Local fare, marketed under a green banner, is already on its way in.

Where I live, there are now restaurants that serve only locally raised chicken and meats, locally grown vegetables and fruit and even locally produced wine. And believe me, southern Ontario is not Burgundy or Tuscany when it comes to regional cuisine. We already know what happens to the Norwegian salmon processed in the Chinese plant. The only fish on the menu in the new breed of upscale restaurant is that caught from Lake Ontario, which you may or may not want to eat, depending upon your tolerance for Great Lakes toxins. Your menu, like the world economy, has just got a lot smaller.

Where will tomorrow's food supply come from when it no longer makes any commercial sense for North Americans to import lamb from New Zealand or for the British to get their lettuce from Spain? It won't be easy to just grow it. Local food production in most places in the developed world has lagged well behind consumption, leaving us increasingly dependent on farms elsewhere in the world. The province of Ontario, for example, imported less than 25 percent of its food in 1988, and now imports over 40 percent. And Ontario is no different from anywhere else in North America. In 1980, the US imported just over 40 percent of its fish. By 2005, it imported 70 percent. And those imported frozen lamb chops have gone from 10 percent of the US market in 1980 to over 40 percent twenty-five years later. Even fruit has gone from 5 percent to 25 percent.

But that is going to change. During the 1990s, the American economy lost two acres of farmland every minute. Tomorrow's farm sector may be regaining those acres at the same pace.

And you may run into your long-lost barista selling heirloom tomatoes at the local farmers' market.

If you believe in markets, you may be surprised by what the future looks like. Not personal spacecraft or gleaming mega-cities—those are the daydreams of the era of cheap energy. The future will look a lot like the past. And that means more farms.

We have already seen that soaring transport costs and the subsequent collapse of commuter traffic will depopulate the suburbs. The farther they are from where people work, the emptier they will get.

Half-built subdivisions are already being abandoned. Without the steady cash flow from new home sales, developers will bail, leaving muddy fields with a few foundations poking out of the ground. And when interest rates finally catch up with where inflation will be going, and financing new subdivisions gets that much more expensive, the corrosion at the edges of the city will only accelerate. Don't be surprised how little time nature needs to reclaim what was only thirty or forty years ago countryside. In Lehigh Acres, an abandoned development in Florida, authorities have shut down more than 100 marijuana grow-ops. Not the most nutritious crop perhaps, but a pretty clear sign that nature and entrepreneurs abhor a vacuum.

In a smaller world, market forces will be on Mother Nature's side for a change.

As soaring transport costs take New Zealand lamb and California blueberries off Toronto menus and grocery-store shelves, the price of locally grown lamb and blueberries will rise. The higher they rise, the more they will encourage people to raise sheep and grow blueberries. Ultimately, the price will rise so high that now unsaleable real estate in the outer suburbs will be converted back into farmland. That new farmland will then help stock the grocery shelves in my supermarket, just like it did thirty or forty years ago.

Where is the food of the future going to come from? Your own backyard. That shift in food supply is already starting to take place. An increasingly carbon-conscious and energy-conscious food consumer in North America is already clamoring for more homegrown food. Not only does replacing foreign food with local food save energy, but in the process it reduces carbon emissions—a double win in an economy that not only has to contend with triple-digit oil prices but that will soon put a price on burning oil as well.

In fact, there are a lot more things to like about local food. As food takes up a bigger and bigger portion of our budget, it makes good sense to keep that spending in your community. Better to have your money help build the local economy of the future than to flow to the bottom line of a distant corporation. And here is the other thing: local, seasonal food is always going to be fresher than imported frozen food. Usually that means it should taste better as well, and be more nutritionally complete (one study in the UK found that potatoes were often in storage for up to six months, which leaches out many of the nutrients).

Every weekend during the summer, my neighborhood park, located no more than two miles from the city's downtown and Canada's financial center, hosts a farmers' market. Nearly half the park is covered with stalls where local farmers sell family farm produce to urban customers. These are the same customers who are increasingly eschewing the foreign food sold in their local supermarkets, and they are willing to pay a hefty premium.

Everything from fresh-grown meat and homemade sausage to fresh vegetables and honey can be bought at Withrow Park every weekend during the summer. It's all locally produced and most is organically grown. Even large jugs of freshly rendered pork lard are available for the asking. I guess eating local doesn't necessarily always mean eating healthy.

These farmers' markets springing up all over metropolitan areas in North America have not gone unnoticed by the large supermarket chains. The largest chain in my city has recently announced that it will boost local produce in its store by 10 percent. That shouldn't be too difficult when, like most supermarkets, only a quarter of the produce on its stores' shelves comes from local sources. It plans to have horseshoe displays featuring local produce in the front of every store so that every customer can see its support for the local, environmentally friendly farmer.

I may be able to put food on my family's table from what I can buy at the weekend food stalls in Withrow Park, but boutique farmers and artisanal bakers and cheesemakers provide only a fraction of the food eaten in the developed world. And if you have ever been to a farmers' market and have seen the fleets of Land Rovers and sleek Volvo wagons heading home with their cargos of organic *cavolo nero* and free-range Berkshire pork, it is probably pretty clear that your local farmer is not going to be able to replace the petroleum-dependent bounty of the global industrial food system any time soon.

Today only the relatively affluent can afford to eat locally. This should tell us that food is going to be more expensive when it is no longer subsidized by cheap energy and global wages.

That is especially alarming when you factor in the cost of actually growing the food. We may think of farms as rustic Edens of placid cows in bucolic pastures and merry chickens pecking away in the barnyard, but behind that green facade is one of the most energy-intensive industries in the world. Large-scale, mechanized commercial farming is just a sophisticated way of turning fossil fuels into food. The fields of Iowa, for example, require energy inputs every year equal to between four and five thousand bombs the size of the one dropped on Nagasaki. And that energy comes from fossil fuels—about five and half gallons per acre.

Back when our food came from the family farm down the road, a farmer got about three calories of food energy back for every calorie of work he invested in his land. Today, now that the work is done by gigantic diesel-powered tractors and combines and trucks, we get one calorie back for every ten invested. Like the Canadian oil sands, this is another example of a diminishing energy rate of return. And that diminishing energy rate of return just gets more daunting with every rise in the price of oil. As we have already seen in the case of ethanol production, soaring fertilizer costs, soaring diesel fuel prices and soaring power prices for irrigation all flow through to the final price of the harvest and the cost of the food on your table.

Energy and food inflation are inextricably linked. While in the US economy that linkage is accentuated by ill-conceived ethanol policies, around the world the relation between food and energy prices is far more basic. As critical inputs to agriculture, there is a direct pass-through from higher fertilizer costs, power requirements or diesel fuel. All of those inputs have risen, whether you measure them per acre, per pound of food or by any other metric you can think of. And the more mechanized agriculture gets, the more energy dependent it gets.

Meanwhile, the world is finding ways to consume more food all the time. Hamburgers are replacing rice bowls just as quickly as Tatas and Cherys are replacing bicycles in China and India. As people raise the protein content in their diets by consuming more meat, pressures are in turn felt on world grain markets to accommodate soaring demand for livestock feed.

Global consumption of meat has grown at double the speed of population growth over recent decades. Currently, we consume about 90 pounds of meat per year on the planet—up from 60 pounds in the 1970s. And most of that rise in meat consumption is coming from the developing world.

It takes a whole lot more grain to produce a pound of meat than it does to make a loaf of bread. In fact, it takes close to seven pounds of feed to produce a pound of beef or pork, and about two and half pounds of feed to produce a pound of chicken.

So when world meat consumption soars, we have to grow a lot more grain than we used to simply to raise the additional meat. That's why demand for oilseeds and grains grew faster than supply in seven of the last eight years. From 2000 to 2007, world grain inventories fell by 50 percent and are now at the lowest levels on record.

That's fine if you are the one who gets to eat the steak, like the soaring urban populations in China and India. But the more beef you graze, the less bread you can make. Grain that might otherwise have gone to feed people now feeds cattle. You are going to like this arrangement a lot less if you are a rural peasant whose diet depends on those very grains you are now feeding some steer. You starve as soaring grain prices suddenly put the staples of your diet out of reach. That's why more and more countries in the third world are either banning or severely limiting exports of grain and other food-stuffs in the face of worsening domestic food shortages.

World Bank president Robert Zoellick warned in October 2008 of a mounting "human crisis" as 44 million of the world's poorest people were driven into malnutrition as a result of high food prices. "While people in the developed world are focused on the financial crisis," Zoellick said, "many forget that a human crisis is rapidly unfolding in developing countries. It is pushing poor people to the brink of survival." The rising price of basic foodstuffs may cut into the budgets of people in the OECD, but in poorer countries it can mean life or death.

There is only so much arable land on the planet. In fact, climate change may mean there is less of it all the time. All of the huge gains in world food production over the last four decades have come as a result of steadily improving yields per acre. Acreage under cultivation has itself not grown.

Meanwhile there are more people all the time, with more energy-intensive appetites. And more cars, which means less oil for the tractors and, as we have seen, less corn for tortillas when it is diverted toward ethanol production.

Peak oil may soon give us peak food.

Back at your local market, there will still be food on the shelves, of course. But it is more likely to have come from nearby, and it is certain to be more expensive. When energy prices are high, the cost of the labor and the fuel to grow food, as well as the cost of shipping and retailing it, will be higher and will rise with every dollar increase in the price of a barrel of oil.

That means that with every rise in the price of oil there will be a little (or a lot) less money leftover in your budget for anything other than food. As a percentage of household expenses, food is already going up all the time.

But as we have seen, the price of anything imported will soon be going up along with the cost of shipping it.

And the price of anything produced domestically is also going up along with the cost of labor.

And the cost of borrowing the money to buy it will also go up as interest rates track inflation.

Suddenly the price you pay for those boutique blue potatoes at the farmers' market is looking like a bargain. As high oil prices work their way through the economy, there is going to be less money to spend, and many things that look like essentials today may look like luxuries tomorrow. My double-shot latte already costs over $3. How much more can I afford to pay for that wake-up call I give my body every morning?

EFFICIENCY AND SPECIALIZATION

The global economy is all about efficiency. It commands us to specialize in what we are the best at, just as Ricardo observed back in the nineteenth century. And the bigger the world, the better we are able to do that. The more we can trade with the rest of the world, the larger the markets that we can access, and the more specialized niches they come to offer and support.

If I make something that only ten people want to buy, I'm going to have a very hard time making a living at it. And I better be charging quite a bit for each thing that I am making. And that means there has to be something very special or unique about it to grant me the market power to charge that much. On the other hand, if a million people want to buy what I make, I could earn a very good living and I won't have to charge that much for it. Now I've got volume on my side. For most products the question of profitability comes down to one thing: market size.

What takes me from a market of hundreds or thousands to a market of millions is going from a local economy to a global economy. And what connects the local economy to the global economy is trade.

The more an economy trades with the rest of the world, the more its producers and consumers become connected to the global market. Global markets make producers exporters and consumers importers.

A global economy encourages us to become more and more expert in increasingly esoteric ways. You can put a lot of eggs in one basket if the basket is big enough. South Koreans wouldn't be much good at making semiconductors if they were producing chips just for Korean computers. But they make semiconductors not only for themselves but for the whole world. In order to specialize so intensively in making semiconductors, Korea doesn't make other things they might enjoy. Coffee, for example. So the Hermit Kingdom imports coffee from countries that don't make semiconductors. And everybody is better off, since no one has to drink Korean coffee and Brazilians don't have to make their own computer chips from scratch.

But what happens to that trend of ever-greater specialization when the broad economic forces that drive globalization start to reverse?

As soaring energy-induced increases in transport costs swing the economic pendulum from the global economy back to the local economy, we suddenly need to become generalists. We have to refocus on smaller, more local markets that may not be large enough to support the specialization that the broader global market previously provided.

Instead of being the super-efficient producer of a specialized widget, I may now have to become more of a jack of all trades by producing five or six different types of widgets for a much smaller regional market. I may not be as efficient producing five or six different types of widgets as I was producing a single widget but, thanks to exploding transoceanic freight rates, I don't have to be.

I'm no longer competing with everyone in the rest of the world to survive. Now I am competing only with other local producers who operate at more or less the same level of efficiency as I do and who pay basically the same wage rate.

Labor unions call it "fair trade," which really means trade with only those who are no more competitive than you are. I call it "local trade."

If triple-digit oil prices suddenly render the cost of shipping what I produce across an ocean untenable, then I am not going to worry about supplying Asian or European markets. I am going to refocus on local markets, where soaring freight costs work to my advantage, not my detriment, by keeping overseas competition from selling in my neighborhood's stores. I may have to withdraw from my competitors' markets, but then my competitors are about to withdraw from mine, driven by exactly the same economics.

Similarly, the practice of just-in-time inventories will no longer be viable with tomorrow's shipping costs. It will become far cheaper to warehouse larger inventories than to continually be shipping product across the ocean. Higher warehousing and other distribution costs will ultimately be passed on to the consumer, just like the increase in shipping costs.

That's why consumer products giant Proctor & Gamble quickly announced a retooling of its supply chain when oil rose to triple-digit levels, shifting from overseas suppliers to local suppliers in close proximity to the company's major markets. Minimizing transport costs, particularly over long-haul transoceanic routes, will be the driving force behind tomorrow's supply-management practices.

It is a reciprocal downsizing of the world for everyone. In retreating to our home markets, all of our horizons are going to get smaller, whether we are North American firms, Japanese firms or European firms.

But shifting from a large to a small market is not like ordering a tall coffee at Starbucks instead of a venti. When you change the size of the cup, you change what is inside.

What drives the efficiency and specialization that come from competitiveness in the global market is technology. Technology is both the cause of these productivity gains and the effect, because technology breeds technology. In fact, whether we are talking about computer memory or processing power or network capacity or just about any other way of measuring technological advance, Moore's Law says that it is going to double every two years. In 1972, Intel's 8008 chip had 2,500 transistors. A Pentium 4 chip has 42 million. In other words, technology grows exponentially.

At least it has for half a century, spurred on by expanding markets and cheap energy. What happens when the rules of the game change?

Smaller markets won't be able to sustain the same level of specialization, which is the lifeblood of global technology. Without the ever-increasing degree of technical specialization seen over the last few decades, it is unlikely that the next ones will be able to produce the same pace of technological progress as we have seen in everything from video games to genomics.

Not incidentally, that Korean microchip has to be manufactured before it can incur shipping costs, and that requires energy as well. A standard 32 MB DRAM chip, which tips the scales at only 2 grams (7 one-hundredths of an ounce), requires a staggering 1.6 kilograms of fossil fuel (and 32 kilograms of water) to make. The embodied energy in that little piece of hardware is about two hundred times higher than what you would find in your car, for example, and a thousand times higher than the energy it takes to make steel from crude iron ore. Needless to say, Korea trades a lot of its precious semiconductors for oil, rather than coffee.

What is at stake here is not whether iPhones and GPS devices for our cars will continue to be cheap and plentiful in the smaller world around the corner, because the important thing about technology is not how it entertains us (though this is the most obvious thing about it), but how it powers our economies. Rising productivity is the mechanism that keeps churning out wealth. The more stuff we get for the same amount of work, the better off we are. And it is technological progress that allows us to get more and more productive—the innovation that will allow my company to use the resources or capital or labor at its disposal more efficiently than my competitor.

And when that happens, whatever it is I make, whether it is coffee or semiconductors, gets a little bit cheaper, leaving everyone with a little more money to spend on something else, and leaving my company with greater profits. When everyone deploys technology in this way (though hopefully not as successfully as I do), the result is economic growth.

But take away the technological growth, and suddenly economic expansion itself begins to look like another artifact from the era of cheap energy. Unless we can find either a way to reduce our energy requirements or an alternative source of energy, this is another way in which peak oil will lead to peak GDP.

STAYING HOME

Menus in restaurants are going to change, but what you will notice most is that you will be looking at them less and less.

When it costs well over a hundred dollars to fill up your tank, you will be driving a lot less, and most people drive to the restaurants they eat at. Either that or the food drives to them. Just as they will keep you out of restaurants, soaring fuel costs will also have you ordering in less, because eventually those fuel charges are going directly onto your delivery bill.

Even though I'll be driving less, filling my tank will still leave me with a lot less discretionary income to spend. I'll eat at home more, buy more locally grown food and, alas, drink less exotic coffees. But energy prices can squeeze food expenditures only so far. On the other hand, they can squeeze nonfood expenditures right out of my budget. If you think triple-digit oil prices crimp my eating style, wait till you see what it does for my future travel plans.

People don't generally realize what gas-guzzlers planes are. It's only been by following the airlines' considerable carbon trail that the public has come to understand just how much fuel is burned in the sky.

Your standard return flight from New York to London burns up about 24,000 gallons of jet fuel or 112 gallons per passenger in a Boeing 767. That's as much as a mid-sized American car burns in three months. And the jet fuel that is powering the plane that is flying you over the Atlantic costs basically the same price as the gasoline you put in your car. Do the math and you start to get the picture.

Air travel just doesn't make sense in a world of triple-digit oil prices. Not as long as jet fuel is made from oil (and the prospects of fueling jets with ethanol or biodiesel are about nil; in addition to the problems associated with using biofuels in cars and trucks, ethanol is insufficiently energy-dense, and biodiesel gels at the low temperatures of high-altitude flight). It takes over forty times as much energy to fly something over the Pacific Ocean as it does to ship it. And even transoceanic shipping isn't that cheap when oil costs $100 a barrel as Chinese steel exporters have already found out.

For the last thirty years, since airline fares were deregulated in the United States in 1978, the inflation-adjusted cost of airline travel has fallen by more than 50 percent. The situation is even more pronounced in the British Isles, where discount airlines like the UK's EasyJet and Ireland's Ryanair advertise fares about as close to free as you can get. And, not surprisingly, the volume of airline travel has grown exponentially as the real cost of flying has fallen steadily. While air travel was once glamorous and exotic, it is now an entitlement we all enjoy—for the time being. That's about to change.

Airlines were quick to pass on crippling fuel-cost increases to customers in the form of ever-greater fuel surcharges on airline tickets. Expect all future oil price increases to go directly onto your fare. But for most airlines in the world today it will be a losing battle to stay abreast of soaring fuel costs even if they ultimately try to pass them on to customers.

Most airlines need to see oil prices below $80 to so much as break even. All face operating losses at triple-digit fuel costs. At over $100-per-barrel fuel costs, US airlines can expect to lose around the $7 billion a year. As one airline executive recently commented, "This is structural, not cyclical."

Unable to immunize themselves from soaring fuel costs, airlines will have to dramatically change business practices. Flights of half-empty planes traveling to more remote secondary locations will be canceled. Connecting to a regional hub like Atlanta or Dallas will now become much more difficult if you don't live in a major city. You may soon find yourself driving past the empty local airport on the way to a bigger one in the next city down the highway. Or, more likely, you will be looking out the window of a bus.

The cost of burning jet fuel will disconnect those smaller locations from airline hubs. And that will be only one of the many ways that soaring energy prices will disconnect the people in those locations from the rest of the world.

Even with aggressive pruning of noneconomical service, the tandem of soaring operating costs and declining ticket sales will force many airlines into bankruptcy, only further exacerbating the reduction in airline service. Even the temporary shock to airline travel following the September 11, 2001, terrorist attacks on New York City and the Pentagon had a dramatic impact on the industry. Four of the six largest American airline firms tipped into bankruptcy. And the energy shock that is hitting us is not a momentary political blip. This is permanent.

Industry consolidation will only further exacerbate the service cutbacks, as cost-conscious airlines refocus on fewer but more profitable routes with much greater load factors to offset mounting fuel costs. And as airline costs are constantly cut, the public is likely to become more concerned about flight safety, only further dampening demand. As if to underline just how sensitive the airlines are to the price of jet fuel, Ryanair gave some justification to everyone who fears flying when it was reported in *The Times* (of London) that management had sent a memo requiring pilots to cut back on the emergency fuel they carried. To save on costs, the newspaper reported, pilots would be allowed to carry only 300 kilograms of extra fuel, or enough to stay in the air for about five minutes.

And if soaring fuel costs weren't enough for struggling airlines to contend with, taxes on carbon-spewing airline travel could soon become the new sin taxes of the 21st century. The emissions per passenger from a round-trip airline flight from New York City to London are as bad as the fuel consumption. If we move to put a price on carbon emissions, air travel may soon be taxed at the same rate as tobacco and alcohol.

As a result, people will travel less frequently, and when they do travel, they will travel closer to home. Britons are going to be a lot less likely to fly to New York to do some shopping, or to slip off to Ibiza for the weekend or to their second homes in France (unless they take the Chunnel). And Americans will vacation a lot closer to home for exactly the same reason. Visiting your girlfriend in Seattle or meeting up with some buddies for a round of golf in Las Vegas may not seem like an extravagance today, but it will tomorrow.

Business travel will be pared back in just the same way, in part because businesses will be much less global in scope and in part because cost control is going to take on new urgency in an economic environment of spiraling energy prices. That colleague or client on the other side of the ocean whom you see regularly for meetings or sales calls is going to become a bit more of a stranger as teleconferencing and email replace face-to-face contact. We will be farther and farther away from parts of the planet that today seem right around the corner.

Put less vacation travel, less business travel and less air freight together, and all those newly built airports, like the new terminal in Toronto or the planned third runway and sixth terminal at London's Heathrow, will soon become gleaming mausoleums to a past age of cheap and abundant energy.

The tourism and recreation industry's ability to weather soaring fuel costs will come down to one factor. As in real estate, location will determine the survival of the fittest in a world where people travel much less than they used to.

Some of the world's hottest vacation destinations could quickly become ice cold. Maybe it's just as well that there will be fewer tourists gawking at the wildebeests in the Serengeti. Or maybe not. When the tourists stop coming and the economy slows down, the poachers start hunting.

But for better or worse, there will be fewer cameras snapping shots of wildlife in Africa. While the wildlife in the Rift Valley in eastern Africa is one of the world's greatest sights, what will it cost to fly from New York to Nairobi? And how many airlines will even be flying there?

Ditto for the Great Barrier Reef and the pristine tropical jungles that my family and I have visited along the Queensland coast in Australia. And what about the cost of that flight to Lima en route to Machu Pichu, perched high in the clouds of the Andes? Those ancient Incan ruins could probably use a break from the millions of feet that now tread those once sacred steps.

But the Peruvian economy is going to miss those tourists. Peru added one million jobs in tourism between 2000 and 2005, and no one will be happy to see them disappear. And while Australia has a much more diversified economy than Peru's, tourism still accounts for about 4 percent of GDP there. That's more than the country counts on from agriculture, for example. Globally, tourism makes up about 10 percent of GDP, and as much as 30 percent in poor countries—hardly a trifling amount. If people stop traveling, there will be a lot less money changing hands and a lot less spent in countries that really need it.

But the $7 trillion the world spends on international tourism each year will not be the only thing we will miss in a smaller world. Go just about anywhere, from a beach in Greece to a Himalayan village in Nepal, and you are likely to find a Canadian, an Australian and a New Zealander sitting at the bar. It may just be that young people from these relatively far-flung places have to make heroic journeys to see the world, but long-distance travel is now a cultural institution in many countries. Just as guesthouse owners in Chiang Mai and *pensione* managers in Pamplona will miss sunburned adventurers and their traveler's checks, so we will miss our own trips abroad and the way seeing new places revitalizes our own cultures.

As our world shrinks, we may retreat to the footsteps of our parents, seeking local as opposed to global escapes when we go on holiday. Are long-abandoned dancehalls in the Catskills about to get a facelift as New Yorkers head upstate to escape the summer heat, instead of biking through Burgundy? Will my kids go on a month-long canoe trip in northern Ontario instead of hiking through Kashmir like I did? Resorts close to large metropolitan centers will survive. Some may actually prosper—more of their home market will be staying close to home when booking vacations. It is not as though the world is going to come to an end if we travel less. But it is definitely going to feel smaller.

STAYING PUT

Getting away from it all means something very different for the citizens of the third world than it does for the people in your neighbourhood. You may dream of unwinding by bone-fishing in the Bahamas or touring the vineyards of Tuscany, but for someone in Algeria or Guatemala, getting away may mean risking one's life in an overcrowded fishing boat or in the back of a sealed transport truck to sneak into Europe or North America to pursue the promise of a better life.

Lenin once remarked that people vote with their feet, and a newly stationary world will have a lasting effect on global migration. The last thirty years of prosperity since the previous oil shocks have been the backdrop to one of the largest migrations in world history. And like most of history's great migrations, the flow of people has been in one direction: from the developing world to the developed world. Whether it's the hordes of migrants from the Middle East and Africa to Europe or from Mexico and the rest of Latin America to the United States, the advanced OECD economies have attracted people from all over the world at a pace seldom seen before.

An obvious factor behind this huge movement of people has been a parallel explosion in the volume and growth of global trade. People and goods have moved together all through history, and there is no reason why they are about to stop now.

Not only have barriers fallen for the movement of goods, so have barriers for the movement of people. Immigration quotas in OECD countries were either raised or ignored as increasing domestic labor scarcities encouraged the importation of cheap labor from abroad. As the jobless rate plunged to new record lows in Europe and America, more and more migrant labor was brought in. And with each tick down in the national unemployment rate, more and more jobs on the lower rungs of the economy went unfilled by an increasingly discerning domestic labor force, leaving immigrants to fill the growing number of vacancies going unfilled in the job market.

In turn, as immigration began to shift the complexion of Western countries, OECD societies became increasingly interracial and multicultural. Large Muslim populations have sprung up throughout Western Europe as a response to the relatively tight labor markets there. Huge numbers of migrant workers have found their way to key European economies like Germany, France and the UK.

At the same time, more and more Spanish is being heard in California and Texas these days. Mexico may have ceded almost a third of its territory to the United States after the Spanish-American War, but its population is slowly taking back through migration what their forefathers lost in battle.

What happens to those massive streams of people once oil is no longer flowing as freely? First, global trade will downshift into a much lower gear, which in and of itself will be a powerful brake on migration. Already, the recession has brought growth in global trade to an abrupt halt. And as higher and higher oil prices take a toll on the growth of our still oil-dependent economies, unemployment rates will rise in most OECD economies. A rising jobless rate means a less discerning domestic labor force and hence fewer unfilled job vacancies in our labor markets.

Tighter competition for scarcer jobs has historically translated into diminishing political support for higher immigration levels, and enthusiasm for new workers is already cooling. In Australia, immigration targets are set to fall in 2010 for the first time since 1997, and Canada has also given notice that it will be welcoming fewer newcomers in 2010. In the UK, the Home Office has raised the bar for applicants arriving without a job from a Bachelor's to a Master's degree. And in the US, rising unemployment and stricter patrols along the Mexican border have slowed illegal immigration to a trickle.

And what does that mean for the developing world? As challenging as triple-digit oil prices will be for the world's richest countries, think how much more challenging they will be for the poorer countries. Just as climate change is already affecting the poorer nations near the equator more cruelly than it does the richer, more temperate countries, rising fuel prices hit those places a lot harder as well. According to the International Energy Agency, every $10 increase in the price of oil causes the countries of sub-Saharan Africa to lose 3 percent of their GDP. What happens to their economies when the global oil market becomes a zero-sum game and the poor countries are unable to keep up with the bidding?

When the developed world starts to tighten its belt, the economies of the developing world get strangled. Everyone, rich and poor, bought into globalization because everyone benefited. Incomes rose globally, and at historically impressive rates. Because we wanted to pay less for shoes, someone in Vietnam moved from the rice fields to a factory. He or she may not have lived in luxury, but then societies don't go from subsistence farming to first-world prosperity in one ambitious step—though, as the growth rates of the BRIC countries and the so-called Asian Tigers show, it is possible to move up the economic ladder in a hurry in a global economy. But rising energy costs will soon decouple that factory in Vietnam from its North American or European market. As North American and European markets return to local sourcing, they will sever their trade links with the developing world and force that world to find another way to grow. As our world becomes smaller, their world becomes poorer.

Take Kenya, for example, which supplies about a third of Europe's cut flowers. Just before Valentine's Day in 2007, environmentalists living in London and Paris called for a boycott of African roses on the grounds that the environmental cost of shipping all those flowers was no way to say "I love you." But that was a real problem for Kenya, where the flower industry employs about half a million people directly, and about a million more indirectly. The Kenyan government estimates that about 14 percent of its citizens are dependent in some way on flower exports. Growing flowers and shipping them to Europe accounts for about 15 percent of Kenya's GDP, a staggering amount if you bear in mind that as important as the automotive industry is in North

America—so important that it seems to be entitled to endless concessions and handouts from the government—it accounts for only 2.2 percent of GDP in Canada and 1.3 percent in the US. A rose picked in Kenya in the morning can be purchased in Europe in the afternoon, a luxury clearly dependent on cheap jet fuel (and cheap carbon). If Europeans start buying fewer flowers, or start growing more of them closer to home, the effect on the Kenyan economy will be crippling—a lot more crippling, for example, than the collapse of the auto industry in North America.

The same can be said for the countries that grow over half of the world's coffee. In Brazil, Vietnam and Colombia, agriculture accounts for more that 20 percent of GDP. Of course, we are not going to start growing coffee in Canada or the UK. But if we import less, the coffee farmers of the developing world, like the flower growers of Kenya and the sheep farmers of New Zealand, will begin to switch to crops with more local markets.

In addition to a tapering off in international trade, the developing world will feel the pinch of slowing economic growth or even contraction. Remember that as long as economic output is a function of the amount of oil you can feed the machine of the economy, peak oil will translate into peak GDP. In a world where global population increases by about 80 million people each year, that means a lower standard of living for everybody. While returning to the lifestyle you knew a few years ago might not be something you and your family would particularly enjoy, you probably would not suffer all that much either. But for someone in the developing world just fighting his or her way out of poverty or even hunger, a step backwards could be a terrible reversal of fortune. If you are one of the 1.4 billion people in the world who live on less than $1.25 a day, there is not a lot of room to cut back.

Not only will the inevitable slowing in global trade cause economic havoc in the developing world, in many of those countries with chronic rates of unemployment and poverty it is likely to shut off a critical safety valve: migration. Not only do migrant workers free up scarce jobs at home, they often send generous remittances to their families back home, which in turn are spent in those cash-starved economies. For a country like the Philippines, which supplies much of the world with its nannies, and where remittances from migrant workers overseas constitute 13 percent of GDP, the loss of those remittances would be a heavy blow to its economy. Across the developing world, closed borders in the richer countries will mean trapped and increasingly desperate populations at home.

And what will the cultural and social implications of a sudden change in migration mean in our own societies? Will we go back to the more homogenous cultures of our countries' pasts, or have our societies indelibly changed?

How long will all the traces of globalization around us remain as the flow of both goods and people reorient to a new world that is local and regional in both scale and direction? Will our societies become more closed not only to different peoples, as immigration tightens, but to different ideas as well? In the space of my lifetime, my city, Toronto, has been transformed from a sleepy hollow of provincial WASP culture to one of the most cosmopolitan cities in the world, where different cultures and races interact every day. There can be no doubt that the arrival of new ideas and new people is part of what make great cities so dynamic, no matter what country they're in.

Most of us in the developed world take great pride in the tolerant, liberal societies we have built. And there is a pretty strong argument to be made that the standard of living we enjoy is at least in part *because* of those values. Certainly, the freedoms of democracy and a market economy foster innovation and growth. But it works the other way as well: we are liberal and tolerant because we are prosperous. When we feel we are doing well, we are inclined to be generous and concerned for those less fortunate than us. As our economies grow, so does our sense of responsibility for others.

But as we have seen, it is energy that has allowed our economies to grow as rapidly and steadily as they have. Take away cheap and abundant oil, and all of a sudden you take away the engine of economic growth. As our economic well-being deteriorates, will we continue to remain committed to those principles of freedom and tolerance?

These are not abstract questions—they challenge some of our most fundamental assumptions about our societies. Economic turbulence has not brought out the best in us in the past. Rising unemployment has often bred xenophobia as we start blaming the economy's ills on immigrants stealing our jobs. It is a pattern so familiar that it seems to run like clockwork: the moment clouds start to gather over a country's economy, the scapegoating of foreigners and immigrants begins. Now, after decades of nearly continuous prosperity, it might seem as though this time around we may do better. It may be that we really have learned from past mistakes and injustices. But when groups like America's Leadership Team for Long Range Population-Immigration-Resource Planning, an umbrella for several anti-immigrant organizations, takes out ad space in well-known liberal publications such as the *New York Times* and *Harper's* magazine, it is clear that there is already an ominous undercurrent of resentment that could manifest itself in ugly ways if things get difficult.

A SMALL NEW WORLD

For decades, the best-funded agencies of the world's richest countries spent untold billions on monitoring and assessing the strength and weaknesses of their archenemy, the Soviet Union. Spies and academics, analysts and strategists devoted their lives to figuring out what the Soviets would do next and what the West could do to counter it. And they constantly reminded us of the danger that the evil empire posed. And then one day, without a word of warning from any of our legions of Sovietologists, a seemingly permanent fixture of the world's geopolitical balance more or less vanished overnight. So much for prediction.

As they say in the investment banking business, "Shit happens." That's just a blunt way of saying that much of what you see when you look around is totally unpredictable. Who would have predicted the burgeoning of computer technology in such a short time? Not Bill Gates, who mused not all that long ago that 64 kilobytes of RAM should be sufficient for just about anyone. Today that's a tiny fraction of what's inside a teenager's phone.

As someone who has done this for a living for two decades, let me assure you that forecasting the future is never easy. But it's even less so when one is talking about a radical split from what we have known in our lifetimes—namely, a world built on cheap energy. The future may be uncertain, but the end is not necessarily near. History is not simply going to run in reverse just because the cheap oil that got us to this point is getting more and more scarce. We can't assume that the trajectory of progress is simply going to turn downwards like Hubbert's curve.

Among the many "unknown unknowns" we should expect are unexpected innovations and challenges. Smart, creative people will come forward with new ideas and products, and technologies that are not yet on the shelf may see some kind of breakthrough that will change the game. Geothermal energy, for example, may surge forward as carbon-intensive sources of electricity generation become prohibitively expensive in a world of carbon tariffs and fossil-energy scarcity. In California alone there are forty-three geothermal plants pumping out 18,000 megawatts of electricity. That is less than 5 percent of the state's electricity demand today, but it is exactly the sort of thing that could change in the world right around the corner. New Zealand too sits atop a wealth of geothermal energy. While New Zealand currently gets only a slightly higher percentage of its electricity from geothermal than California, that too could change in a hurry. Australia has no installed geothermal capacity, and that should change in an even bigger hurry.

There are other plans out there, from generating biofuels from oil-secreting algae to storing solar energy in molten salt in North Africa. No one is suggesting that we don't have any tricks up our sleeves, or even that the best tricks aren't somewhere not yet up our sleeves. And of course, many of the solutions we have at our disposal are not tricks at all. We have wind turbine and photovoltaic solar panel technology sitting on the shelf. We don't have the turbines or the panels themselves, since they are on back order from most of the world's factories, but that is about all the evidence you need that there is a great opportunity there for entrepreneurs. It is no easy thing to turn a car plant into a wind-turbine factory, but then, Detroit turned car factories into bomber and tank plants pretty much overnight in 1942. Just because something isn't easy doesn't mean it is not going to get done.

While undoubtedly triple-digit oil prices will light the pathway to finding new fuels in the future, simply learning to use less energy will be the answer in the here and now. We know we can suppress our oil appetite when our economy is shrinking in a recession, but that's little solace if we want to grow. The challenge, of course, will be to expand GDP without burning more oil. It is much the same challenge as that posed by the Kyoto Accord, only this time it's not United Nations bureaucrats but the world's depleting oil supply that has thrown down the gauntlet. The silver lining is that in finding a solution to the energy challenge, we will simultaneously find solutions to our carbon challenge. They are, after all, sides of the same coin.

There is certainly still a lot of low-hanging fruit to reap in terms of energy efficiency out there. Air travel, suburban sprawl, SUVs, energy-sucking plasma TVs, frozen lamb chops from New Zealand, you name it—they all are going to be up for grabs in the new world around the corner. Cutting back even a fraction of our energy consumption would have a huge global impact if everyone did it. But as we learned from the age-old efficiency paradox, we must ensure that efficiency leads to true conservation, and not to a rebound in energy demand.

The fact that it takes 50 percent less oil to produce a dollar's worth of GDP in the American economy than it did three decades ago has not stood in the way of consuming ever-increasing amounts of oil. With both global depletion and OPEC's cannibalization of its own export capacity limiting future oil supply, neither America nor any other Western country will be able to do that again. That in itself is a glimpse of how different the future will be from the past. Increasing the productivity of oil by another 50 percent over the next three decades isn't going to leave very much room for economic growth if the amount of oil we consume is fixed or, even worse, declining. In fact, we may need those kinds of efficiency gains just to offset what we are likely to lose in supply.

When over half of the oil we consume is burned as motor fuel, the process of weaning our economy off oil has to start on the road. No doubt in the future people will find another way of moving themselves and their goods around the world without burning oil, but even if everything goes according to plan (and very little these days does in Detroit), we may have 100,000 electric-powered cars on the road in America by 2012. Reducing the oil consumption of the other 247 million gas-guzzlers means somebody has to get right off the road. A future where gasoline will cost as much as $7 per gallon will see many of us taking the exit lane. But will there be a bus or subway to get on when we do?

Looking down the pipe and seeing one of the largest fiscal stimulus packages ever on the way, we can also see hopeful parallels with the past. Just as Lucius Clay directed billions of dollars of taxpayers' money in the 1950s into the massive expansion of roadways and interstate highways that created a new way of life for America and, later, the other big developed countries, so President Obama now has the opportunity to do the same for an economy that must soon adapt to triple-digit oil prices. Will the billions earmarked for infrastructure be an investment in our past—in more highways and in an obsolete auto industry—or in the future, in public transit and electric cars? Already, Obama has indicated that future energy policy will be inextricably tied to carbon policy, underscoring the new reality that it won't just cost money to buy oil but to burn it as well. From overturning the previous administration's blanket approvals for offshore drilling, to promising to force coal utilities to pay for their emissions, to allowing individual states like California to enforce their own automotive efficiency standards, Obama has signalled a sea change is at hand.

Bankruptcy and collapse in Detroit or in similar industries in other sectors and other countries doesn't mean that the factories will disappear or that the engineers or workers will suddenly forget how to do what they were trained to do. A little "creative destruction" might be just what is called for. Although we will be increasingly energy-constrained in the future, it is not as though we are going to stop building things—even cars—and there will be people around who know how to go about it. If the players in today's market can't figure out where to go from here, the next generation of innovators certainly will. It may be companies like the ZENN car, companies that are knocking on the door right now. Or it might be someone with a new idea we haven't dreamed of yet. But figuring out how to get the most out of what we have at our disposal is going to be the key to adapting to a smaller world, and that applies to assets like infrastructure and trained workforces more than anything else.

Not that any of this is going to be a snap. Sure, the US economy managed to make a lot of tanks and planes in the Second World War. But they had to stop making cars for a while. Almost anything we roll up our sleeves to do is going to mean not doing something else. One thing that is going to be especially difficult is going down the list of options and seeing what we can afford. The credit crisis of 2008 gulped down a lot of wealth that would have been very useful to have around in the not-too-distant future when it comes time to adjust to both the challenges we foresee and those we haven't yet. Leaders around the world like Barack Obama in the US and Gordon Brown in the UK have made public their intentions to cut carbon emissions by 80 percent by 2050. Though environmentalists say that is not going to be enough to get the job done, it is still going to cost a lot of money. One US study figured it would cost about $50 per ton to cut emissions. When you take into account that we will need to cut several billion tons each year, you see that there is going to be less money to go around for other things.

What is left in the coffers and what use we make of it will help determine what our smaller world will look like. Will we spend our last dollars investing in new rail systems or refurbishing (or expanding) crumbling roads and bridges that are probably doomed to some form of abandonment anyway? Will we build new airport runways and terminals or develop the infrastructure necessary to move freight by boat on our lakes and rivers? Will we use our municipal taxes to extend sewers and power lines out into the sprawling suburbs or will we invest that money in redeveloping our dense urban centers? We now face the choice between propping up a collapsing way of life based on car-dependent suburbs and designing and building systems better scaled to the future we face.

Change is coming whether we like it or not. And for all the unknowns just beyond the horizon, there are more than enough knowns we can start planning for. We know that projected oil supply won't meet projected demand, and we know that for exactly this reason, both supply and demand are in for a turbulent ride. We know that the cheap-energy economy we have grown so comfortable with is tapering off. The question is whether we are prepared to handle it. The sooner we see it coming, the better we can not only adapt to it but benefit from its effects. And we may find many silver linings tomorrow from the sacrifices we make today.

Instead of smog-choked cities ringed with suburbs and crisscrossed with highways, we have an excellent shot at finding ourselves inhabiting smaller-scaled, walkable neighborhoods and small towns built (or rebuilt) to suit the small new world. Instead of far-flung Levittowns, today's suburbanites may be living in rejuvenated city cores or redeveloped commuter villages linked by rail or public transit or even rivers. Development always follows the transportation routes, just as water follows the path of least resistance. Build the transportation you want, and you won't have to wait long to get the kind of town suited to the future.

One thing these neighborhoods will have in common is a viable relationship to the farms that feed them. Small cities and commuter villages will be nodes of local agricultural production. Larger cities will no doubt boast a lot more fruit trees and vegetable gardens (Vancouver, British Columbia, has already changed its bylaws to encourage urban beekeeping, and you'll find beehives on the rooftops of Paris), but the real change might come in the form of vertical farms, or "farmscrapers": multistory greenhouses (and perhaps cattle barns) that produce food all year under controlled conditions. No crop failures, no need for chemical pesticides, no need for extensive transportation networks, and a great way to make use of recycled, or "gray," water. If you happen to live in a city in the American Southwest or Australia, where drought threatens not only your golf greens but your very way of life, an efficient way to turn scarce water and abundant sunlight into food is going to be very important indeed. If we can build multistory parking lots to leave our cars in while we are at work, it is not exactly science fiction to think we can put up multistory buildings to grow romaine lettuce. Especially if we are no longer driving our cars.

In any case, the return of local food in a smaller world will mean the return of local flavor. McDonald's wants everyone eating burgers, from Nairobi to New York. Coke wants everyone drinking pop, from Atlanta to Athens. KFC wants everyone licking their fingers, from Louisville to Bangkok. Starbucks, for all its lip service to cultural diversity, seeks a homogenized consumer who dutifully orders his latte without even noticing what country he is in.

In short, the global economy strives to create a uniform set of consumer choices and preferences that holds anywhere around the world. The global homogenization of cultures and tastes allows one marketing message to resonate clearly around the world no matter how many different languages it is delivered in. That makes it a whole lot easier on producers. All they have to do is find the place with the cheapest labor force and start building factories. It doesn't matter how far away the customers you are serving may be. After all, in the global economy, everybody buys the same things.

Mass-market size and standardized global commodities reinforce themselves, but at the expense of local tastes in local markets. While globalization brought the Atlantic salmon to dinner tables around the world, it also took away local flavor and context. In a smaller world, all the things that make food worth enjoying will again become more important.

And that will be true of many things. Local tastes and local customs, seemingly headed for extinction in the face of globalization's onslaught, will get a last-minute reprieve. As production returns to local roots, what is produced begins to assume more and more of a local flavor—perhaps not steel plants, but certainly factories that make final goods that are bought by the consumer will once again begin to produce to please their local customers' preferences and tastes. As businesses make that shift, they will reconnect with their long-forgotten and ignored local communities.

Where your factories are located won't just be about where your labor costs rank on some global cost index. You are no longer going toe to toe with some sweatshop in East Asia producing on a massive scale for the whole world market. Instead, it will be about your ability to produce goods that are specific to local tastes and customs. Because, in a world of triple-digit oil prices, your close proximity to those tastes and customs will be your company's single greatest source of comparative advantage.

And if transportation and carbon costs don't bring that factory home, it is likely to be swept back to where it came from on a rising tide of protectionism. Governments around the world, led by the US, are insisting that taxpayers' money be spent on locally produced goods.

As government spending becomes a bigger part of the economy, its bias towards local procurement will shut imports out of more and more of the market place. And once the United States slaps "Buy American" restrictions on its economic stimulus packages, as it has already done on everything from water and sewage projects to bridge repairs, expect Europe, Japan, Canada, Australia, and just about everybody else to follow suit. With the recession claiming as many as 50 million jobs around the world, according to the latest estimates from the International Labor Organization (ILO), there will be a groundswell of public pressure for governments to save local jobs. And that means they will soon be turning their backs on global trade and returning to their local economic roots.

One day soon, you won't be wearing the same thing as your doppelgänger in Tokyo. And, eventually, your house won't look the same as every other house on the continent. The things that made your local environment distinctive will come back to the fore. In the none-too-distant future, the past will spring back to life.

We have already seen that the suburbs will slowly (or quickly) empty out, perhaps shrinking back to the villages they had gobbled up, which will once again be surrounded by farmland. Before that happens, the orderly avenues and cul-de-sacs of suburbia will no doubt go through a wrenching transformation during which they will first be slums and then a gigantic salvage operation, as all those (increasingly scarce) building materials get put to better use rebuilding and retrofitting urban homes. In the abandoned exurb of Fort Myers, Florida, thieves are stripping air conditioners for parts and selling them as scrap metal. Once that process has run its course, things will be built once again with local materials, rather than the uniform, prefabricated products available at big-box home-renovation stores around the world.

Not that there won't be use for handymen in the small new world. I suspect we will all get a lot better at fixing things when they break rather than buying new ones. Many of us may finally come to understand how a toaster works. We will make do with things that have been patched and repaired, and things that are functional rather than beautiful. Our clothes may not be as glamorous, I am sorry to say. But sewing is sure to make a comeback, along with gardening and cooking, though I am not sure I am ready to bet on beekeeping.

Of course, not many of us are likely to master all these forgotten skills. How many people on the subway you take each morning would know how to grow a tomato or darn a sock (or even know what "darning" is)? Not many. But one of the things that emerged in the wake of the collapse of the Soviet Union was the strengthening of networks of friends and family. To get through those dark days, people had to help each other out. Hopefully, we will manage to do the same.

How all this will affect our culture is a topic for others to explore in depth, but it is bound to be a radical change. For many of us today, the local world is largely irrelevant. Just as we no longer eat locally, we dream of spending our time elsewhere, whether for business or pleasure. Many of us have more in common with our peers elsewhere around the world than with our neighbors next door. That's about to change. As scarce oil starts to make your world smaller, you will soon be spending much more time talking to your neighbor and much less time flying around the world. And as that happens you will find yourself worrying less and less about the world's problems and more and more about local concerns. We will soon become far more attentive custodians of our own little worlds. And that is likely to make our neighborhoods better places to live.

But will you be able to get a decent cup of coffee? It may not be as easy or as cheap in a smaller world, and a lot will depend on the choices we make as our petroleum-powered world lurches through change after change. But our ancestors were sipping coffee as they discussed the new ideas that brought this world into being, and that was before anyone thought of fueling a ship or a tractor with fossil fuels. There is no reason we shouldn't be able to fuel a new set of ideas with coffee as well. We will certainly need a jolt of energy.

CHASING THE INCONNU

IT'S MIDNIGHT, BUT THE SUN STILL CASTS luminous shades of magenta and periwinkle off the surrounding mountains. In mid-July in the high Yukon, the sun never actually sets. It just becomes dusk for about five hours before the sun rises again at around four o'clock in the morning.

I am casting into the confluence of where Ptarmigan Creek, a raging torrent coming off the surrounding mountains, empties into the clear, frigid 40-degree-Fahrenheit waters of Pelly Lake. Through its outlet, the Pelly River, the lake connects to the Yukon River and its over 2,000-mile route to the Bering Sea in Alaska. It is that connection that has brought me here on this Arctic midnight, for it's the migratory route of the elusive inconnu.

Technically, I am told, the inconnu is a whitefish, but one that neither looks nor behaves like any other whitefish in the world. It is the only predatory variety of the species and looks more like a tarpon than like any whitefish I have ever seen. Unlike tarpon, however, it seeks the cold, crisp waters of the Yukon and McKenzie Rivers' watersheds and spends the majority of its life in freshwater.

Like salmon and steelhead trout, inconnu are a migratory fish that are sea-bound. However, unlike the migratory patterns of salmon and steelhead, which have attracted voluminous research, little is known of the precise habits of the inconnu, other than that at some point in their life cycle they swim out to the vast North Pacific before returning to their local spawning beds.

What is known is that the inconnu run in the Yukon River system is the longest migratory fish run in North America. From its source, Bennett Lake, the river runs over 1,100 miles to the Alaskan border, draining a watershed that extends over half of the entire Yukon Territory. From the Alaskan border, it runs another 1,200-plus miles to finally empty into the Bering Sea.

The fish were once so abundant that the first explorers of the region, led by Alexander McKenzie and his French Canadian voyageurs, largely subsisted on them. (The fish's name, *inconnu*, means "unknown" in French.) But over time, like so many other overfished species, their numbers collapsed. While still caught in Alaska, where they become sea-bound, they have become a rare and very elusive species in the Pelly River system in the Yukon from where they originate.

I am staying at the remote Inconnu Lodge, in the heart of the Yukon Territory. The lodge consists of five separate red cedar cabins and a 6,500-square-foot main building also constructed of cedar. The main building includes a spacious lounge, a commercial kitchen, an elegant dining room, a tackle shop and a conference center. The lounge is adorned with trophy fish as well as mounted moose, grizzly and caribou, all taken from the surrounding forests. The plush comfort of the lodge stands in sharp contrast to the harsh beauty of its environs.

Across McEvoy Lake, where the lodge is located, there is still snow in July on the top of the mountains that face me. Beneath them, alpine spruce forests interlaced with pink blooming fire-weed are teeming with uncountable hordes of mosquitoes. They thrive in the moisture that comes with daily rain showers in the summer and the immense amount of water available for breeding from the vast interlocking network of bogs, lakes and river systems that criss-cross the territory.

The juxtaposition of snow and mosquitoes within several thousand feet of each other is only one of the many contrasts that make this land so magical. The nearly twenty-four-hour daylight north of the 60th parallel is another, which finds its counterpoint in the few hours of sunlight that the region gets during the winter. But that's a time, of course, when the fishermen don't come and the lodge staff retreat to much warmer and sunnier climates.

Perhaps the most brilliant time of all is in late August and September, when the blue, green and red auras of the northern lights start dancing in the night skies. Caused by particles from solar wind hitting the earth's atmosphere, the northern lights tap into a whole other tourist market even when the fall ice brings the fishing season to an all-too-early close.

As I sit outside my cabin reflecting on the wonder of what I am seeing, I can't help but ponder how much longer I'll be able to come to places like this.

To get here, we have flown 180 miles over some of the most remote mountains in the world from Whitehorse, the 23,000-person capital city of the territory, located in the middle of nowhere. Whitehorse is the center of civilization in a region larger than the entire state of California.

The town is named for the rapids that once stood there, and for the appearance in the churning waters of the manes of galloping white horses. Over 3,000 US Army forces were stationed in Whitehorse during the Second World War to construct the

Alaska Highway, which was rapidly cut out of the Canadian bush in anticipation of a Japanese invasion of the Alaskan coast. It was the first paved road in the entire Yukon.

In total there are 33,000 people who live in the Yukon, of whom only 10,000 reside outside of the capital. By comparison, the territory houses an estimated 250,000 caribou. That's 7.5 caribou for every permanent resident. If I am right about where the price of oil is going, that ratio will soon be going up.

Everything in this lodge has had to be flown in by a 1957 DeHavilland Beaver, the workhorse pontoon plane that literally built the Canadian North. Unfortunately, while it is a true and tested staple of the Arctic hinterland, the Beaver is a fuel pig. Its engine was designed in a world in which oil cost less than 25 cents a gallon. It burns roughly 23 gallons of aviation fuel an hour. The last Beaver came off the production line in 1961. Yet nearly fifty years later, it is being asked to perform critical oil-consuming functions in remote parts of the world where oil now costs forty times as much.

Next to the Beaver as the region's transport is a Hughes 500 helicopter. It can whiz over the Yukon landscape at 125 miles per hour. While much newer than the Beaver, its fuel economy is not that much better.

Yet everything here depends critically on that imported energy. The fly-in lakes boiling with record trout, grayling, northern pike and, hopefully, the elusive inconnu are accessible only through aviation fuel. And once you land at these lakes, the motors of the boats that are tied to the deserted docks run on gasoline. Without the gasoline, there are no motorboats to take you to the fishing spots scattered miles apart across huge Yukon lakes.

Without fossil fuels, no one would be fishing here, and I certainly wouldn't be writing this book here. My laptop is plugged into the electrical socket in my beautifully finished cedar cabin. But the electricity that flows through that wall socket doesn't come from an electrical grid. The nearest grid is hundreds of miles away. Instead, the electrical power that flows through the wall socket of my cabin comes from a diesel generator behind the lodge.

And what about the electric space heater in the cabin? Daytime temperatures around here can push 75 degrees, but there is a reason why there is still snow in the middle of July on the mountains staring back at me from across the lake.

This is the true north. Don't let the midnight sun here fool you. It gets cold at night, even in the middle of July. The same electricity that is allowing me to use my computer powers those space heaters that I turn on every night to keep the cabin warm. Burning ever-more-expensive oil, of course, is what produces that electricity.

It takes an enormous amount of energy every day to run this operation in the heart of the Yukon wilderness. The venerable old Beaver sucks up as much as over three barrels of oil a day. Throw in the helicopter, which some days does the nearly 400-mile round trip to the breathtaking Virginia Falls on the South Nahanni River in the neighboring Northwest Territories, and you are easily talking another one and a half barrels per day. That's four and half barrels a day in aviation fuel alone. Lakes like Jim Cook Lake, where hauling out monster pike and lake trout is routine, are almost 100 miles from the lodge. A day trip there and back like the one we did yesterday burns 48 gallons of aviation fuel, or over a barrel of oil.

And then there is at least another half a barrel (21 gallons) of oil burned by all the outboard motors on the boats used for the fishing trips. Another barrel of oil is consumed in the diesel-fueled generator that provides the lodge with all its power. Add it all up and the lodge burns an astounding six barrels, or the equivalent of over 250 gallons of aviation fuel, gasoline or diesel. Without all that fuel, the lodge simply could not survive.

It's a long way from Cushing, Oklahoma, where the benchmark crude, West Texas Intermediate, is priced, to McEvoy Lake, Yukon Territory. Fuel must be trucked in along a 70-mile dirt track from Watson Lake on the Yukon–British Columbia border, a seven-hour adventure that only the hardiest of four-wheel drive vehicles can make. From there it has to be flown in by either the lodge's Beaver plane or by its helicopter, both of which burn up copious amounts of aviation fuel every second they are in the air.

Loaded with the transport costs of getting here, gas doesn't cost the $4 per gallon most Americans were paying at the pump that summer. Gasoline already cost the equivalent of well over $5 per gallon in Canada then, thanks to much higher federal and provincial excise taxes than in the US. But the lodge must also pay roughly another $5 per barrel simply to bring the fuel into the lodge. Including transit costs, fuel costs for the lodge were now pushing $10 per gallon.

At that price, which is the all-in cost of getting diesel or aviation fuel to the lodge, the daily fuel bill is over $2,500. And that cost will keep climbing as the price of West Texas Intermediate keeps going up in ever-tightening world oil markets. There are 42 gallons in a standard barrel of oil. So a barrel of oil that costs $140 in Oklahoma, will cost $420 per barrel at the Inconnu Lodge in the Yukon.

At these energy costs, will I ever again be able to fish these trophy waters?

The question is particularly apt for the lodge's target market, which is the United States, not Canada. The Canadian dollar, which was only five years ago down to an all-time low of almost 60 cents against the greenback, had risen to par with its bigger brother. And that too was largely because of the price of oil and the growing dependence of American energy consumption on Canadian oil.

And while the cost fallout from triple-digit oil prices is squeezing the lodge operator's margin, it is also squeezing the businesses of the lodge's clients. How will the economic downturn caused by soaring oil prices affect the demand for such places as these?

Already there are changes afoot. No one has to lecture Warren LaFave, the wily owner of Inconnu Lodge and a thirty-one-year veteran of the Yukon resort business, about lodge economics. He knows them cold. His family has operated fishing lodges in British Columbia since the 1930s, and he himself has owned and operated three previous resorts in the Yukon.

His season is barely three months long and he can accommodate only twelve guests at a time at the fishing lodge. Guests turn over every five days, as the plane from Whitehorse comes to fetch the exhausted fishermen and bring the next load in. Over an entire summer season, he has fewer than two hundred spots to fill.

As it turns out, Warren's wife, Anita, went to the same junior high school as I did, decades ago back in Toronto. It's a long way from Wilson Heights Junior High School to the interior of the Yukon, and like so many migrations over the last three decades, it is one that relied on an era of cheap energy. How many more pioneers from the urbanized south will head north in the future, when triple-digit oil prices will make the Yukon and the rest of the Canadian North as remote as it ever has been?

Warren has already announced that he will only operate the fishing lodge business every other year. At least for now, I can still come back in 2010. But he also tells me there are seven salmon-fishing lodges for sale on one river alone in Alaska.

Basic economics tells me that if seven fishing lodges suddenly come on the market at the same time, the market clearing price for an Alaskan fishing lodge is going to be falling. In short, there are more sellers than buyers of Alaskan fishing lodges. And for good reason. Who is going to come in and buy what are hugely oil-intensive businesses when the price of delivering oil to the lodge amounts to as much as $420 per barrel?

At the same time, our lodge owner is also an astute business-man. To survive out here you have to be. The cost of doing just about anything out here leaves very little margin for error. If you can't manage your costs, they will blow you away in no time.

Fortunately, his fishing lodge provides only one of several cash flows that go to paying the bills. He has less luxurious cabins in the back that he rents out to the teams of geologists that have recently been swarming the region. Everything from lead to tungsten is being surveyed in territory that was picked over once already, in the great Klondike gold rush over a century ago, when prospectors had to haul backbreaking supply loads up the gruel-ing 33-mile trail over the unforgiving Chilkoot Pass. Today, snowmobiles have replaced dog teams just as Warren's heli-copter and Beaver have replaced tired feet. But tired feet and dogsleds don't require oil, while snowmobiles and helicopters don't run without it.

More importantly, Warren flies the geologists out in his chopper to where they are conducting their survey work every morning and picks them up at night, for $1,100 per hour.

He also provides an avionics service to the relatively nearby World Heritage site at Nahanni National Park in the neigh-boring Northwest Territories. It's a three-hour plane ride there and back from the lodge and carries a $2,000-plus price tag.

Seventeen years ago, my wife and I canoed down the incredible South Nahanni River as it wound its way from Virginia Falls, twice the height of Niagara, and through four canyons that rival those of the Grand Canyon in Arizona.

Back then, you might see the odd canoe pass you on the river during a week-long paddle from Virginia Falls to the splits, where the South Nahanni rushes out of the McKenzie Mountains to meet the broad-flowing Liard River on the arctic plain. Today, there is a never-ending stream of canoeists doing the paddle, supplied by no fewer than three weekly flights from Frankfurt, Germany, to Whitehorse, Yukon. There is as much Deutsch heard amid the thunder of Virginia Falls as there is English—at least for now. But as fuel surcharges massively inflate the cost of transoceanic travel, there may soon be less and less Deutsch heard in these parts.

You don't survive thirty-one years in the Yukon wilderness without an ability to adapt. And like the weather, things are always changing out here. Warren is already drafting his alternative energy plan.

Conservation is, after all, a central theme of the lodge. All the fishing here is strictly catch-and-release with either flies or barbless spinning lures. That's absolutely essential to maintaining a trophy fishery in the north.

With waters ice-free for only three to four months of the year, fish grow very slowly up here. On average, they put on only about half a pound a year. The lodge's record lake trout is 48 pounds, which in these waters takes almost a century. Even the 33-pound record inconnu taken at the lodge is over sixty years old. Start taking fish out of these waters and it takes nature an awfully long time to put them back.

There is a raging stream that drops 300 feet just in back of the lodge. For $375,000, Warren can build in a turbine and generate a megawatt of hydroelectric power that will save at least a barrel of diesel fuel that he burns in his generator every day for power. It will also save him the fifty tanks of propane that he goes through every summer to heat the hot-water tanks in each cabin for our showers.

He can do something about the boat motors as well. Many are fuel-inefficient two-stroke motors that can be replaced either by more efficient four-stroke motors or, even better, by electric motors. But all of these energy savings that will lower his operating costs and restore his operating margin come at the expense of relatively large capital outlays. And these large capital outlays may be required at a time when a recession softens the demand for high-end fishing lodges.

I don't know how it will all play out. The lodge's cash flows are probably sufficiently diversified that he might be able to operate with a scaled-down aviation operation for geologists and others who simply wish to explore this beautiful region.

Is running a fishing lodge 180 miles away from Whitehorse in the heart of the beautiful but remote Yukon Territory a Canadian version of Ski Dubai? Certainly it's a much scaled-down version. Catching a fish doesn't require the energy of a month's driving, as a day carving turns on the Dubai indoor slopes does. But still, six barrels a day of oil is a lot of fuel in a world of triple-digit oil prices.

If the fishing operation is closed, I along with everyone else will have lost access to these waters. While I caught my fair share of fish here, including a 42-inch great northern pike and an almost 20-pound lake trout, I have yet to snag an inconnu. If oil prices go where I think they're going, I may never get another chance. Which is probably just fine with the inconnu.

———

It is a long way from Inconnu Lodge to my office in downtown Toronto, just as it must seem a world away from where you live and work.

But the world we live in is every bit as precarious as the outposts of the far north, because it is every bit as dependent on energy. Though our skyscrapers and bustling streets, our busy airports and shopping malls seem invulnerably permanent, they are anything but. Like any tired fisherman looking out the window of his cabin at the brooding mountains of the Yukon, you and I depend on a steady flow of oil to keep us safe and warm and well fed, to say nothing of the less tangible things that ensure our satisfaction with life. Take away our energy, and we are all vulnerable.

It is difficult and expensive to get fuel up to the Yukon, but it is becoming more difficult and expensive to get fuel to your local gas station as well. Warren LaFave can just go out and buy the gasoline and diesel he needs. For the oil companies, it is not so easy. They have to find it and produce it, and as we have seen, they are discovering that that is no longer as easy as it once was. And it gets more challenging and expensive every day.

Moreover, the lodge operator can pass on his costs to the men and women who can afford to pay them. For him, the danger is not running out of oil—it is running out of people with deep enough pockets to pay for it. When he runs out of such people, he will have to shut down his little outpost of civilization in a harsh but beautiful world.

The problem is the same for the whole of our civilization. We are not going to wake up tomorrow and find that the world's oil wells have run dry. But we will face the dawning realization that there is a little less each day, and that what remains will cost more to burn.

Eventually, we will face the wrenching choice between adapting to the realities of a new, smaller world or clinging to the artifacts of an old world we can no longer have. On the one hand is a set of costly and risky investments. On the other is a concession of defeat.

I'm willing to bet that we won't go for the second option. We may be energy poor, but we are innovation rich and necessity is the mother of invention. I wouldn't write our economies off just yet.

Yet our world getting smaller isn't just about the global economy and its flow of goods and people, but also about our own life experiences. A smaller world is a less-traveled world, a less-known world, and ultimately a less-sought-after world.

It is a world anchored in local significance and local custom. It is a highly differentiated world, atomistic in structure and diverse in nature. It is a world where our self-identity is defined as much by what we are as by what we are not.

It is a return to a brave new world that has become much bigger again, and one in which we have become a lot smaller.

[ACKNOWLEDGMENTS]

Many thanks to my long-time friend and colleague, Peter Buchanan, Senior Economist at CIBC World Markets, for all his help over the years in our joint study of world oil demand and supply, including our analysis of the growing trend toward the cannibalization of export capacity among OPEC producers. I would also like to thank Benjamin Tal, also a Senior Economist in the Economics Department at CIBC World Markets, for his contribution to our work on how transport costs and carbon pricing could reshape world trade patterns and reverse recent off-shoring trends. Thanks to Kevin Dove of CIBC World Markets for so effectively marketing our research to a broad global audience. Nick Garrison, my editor at Random House Canada, did a great job in providing texture for the book. And last but by no means least, thanks to Colin Campbell for many years ago showing me another way of looking at world oil supply.

INTRODUCTION: **REDEFINING RECOVERY**

p. 2: For an interesting account of the fate of an Atlantic salmon caught off the coast of Norway, see "From Ocean to Plate: A Posthumous Migration," by Sarah Murray, in the November/December 2007 issue of *Orion* (http://www.orionmagazine.org/index.php/articles/article/489/).

p. 3: Highway 401, where it cuts through Toronto, is generally considered the busiest highway in North America, though there seem to be different ways of measuring busy-ness—the section of the I-5 that travels through Santa Monica also seems to have a claim on the top spot.

p. 8: My first acquaintance with the Hubbert Curve came through reading Colin Campbell's *The Coming Oil Crisis*, published by Multi-science Publishing Company, in 1997. A more up-to-date anthology of Campbell's analysis of oil depletion can be found in his more recent *The Essence of Oil and Gas Depletion*, published by Multi-Science Publishing Company, 2002.

p. 11: For an example of the early bullishness that greeted the announcement of Tupi's discovery, see "Brazil, the New Oil Superpower" (November 19, 2007) (http://www.businessweek.com/bwdaily/dnflash/content/nov2007/db20071115_045316.htm). Early estimates had Tupi producing 1 million barrels per day before 2020. More recent estimates put that number closer to 400,000 barrels per day, or less than half a percent of world production (http://www.theoildrum.com/node/3269).

p. 12: M. King Hubbert's reputation has certainly been burnished in the past few years, but it is no exaggeration to say that he was very much an outsider a few decades ago, or that his prediction of peaking US production in his paper "Nuclear Energy and the Fossil Fuels" made him a number of enemies. For an account of Hubbert's career and his struggle to make the point that infinite growth is not possible in a world of finite resources, see chapter two of *The Last Oil Shock*, by David Strahan, published by John Murray Books, 2007.

p. 15: As hard as it is to believe now, there was widespread belief in January 2008, that oil only broke through the $100-per-barrel barrier because a lone trader was having a bit of a lark (http://news.bbc.co.uk/ 2/hi/business/7169543.stm).

CHAPTER 1: **SUPPLY SHIFT**

p. 28: The IEA's "World Energy Outlook" revises the agency's estimate of global decline rate from 3.7 to 6.7 percent—an extraordinary leap with huge implications. It turned out that the earlier estimate had been based on educated guesswork; the higher number was arrived at after a rigorous study of the world's 800 largest fields. Now even the optimists are predicting a production plateau in about a decade. The 2008 WEO is available at http://www.worldenergyoutlook.org/2008.asp.

p. 28: The figures for per capita oil consumption around the world are based on data from the CIA World Factbook (https://www.cia.gov/ library/publications/the-world-factbook/).

p. 35: Though it is hard to believe that drilling for oil in the North Sea is more expensive than putting a man on the moon, that is the way the numbers add up. A 1975 article in *Time* estimates that the investment in North Sea oil infrastructure up to 1980 would cost $11 billion more than the American lunar program (http://www.time.com/ time/magazine/article/0,9171,913489,00.html).

p. 36: An oilfield's output does not follow a smooth curve, which is why the North Sea's peak monthly production arrived fourteen years before its peak annual number. It happened that 1985 saw a particularly good month (https://www.og.berr.gov.uk/pprs/full_production/monthly+oil+ production/0.htm), and 1999 the best year on record (https://www.og. berr.gov.uk/pprs/full_production/annual+oil+production+sorted+by +field+m3/0.htm) according to the UK Department of Energy and Climate Change. But the most important fact is that the North Sea is depleting at 15 percent a year—much faster than the world's average rate.

p. 37: Cambridge Energy Research Associates (CERA) is a bulwark of the energy optimists, and the source of a great deal of the research that comes out to challenge the idea that global oil production is headed towards an imminent peak. Its founder, Daniel Yergin, is the author of *The Prize*, a Pulitzer-winning history of the oil business. In January 2008, CERA estimated that global depletion was proceeding at 4.5 percent—considerably slower than the 6.7 percent the experts at the IEA projected in the 2008 "World Energy Outlook."

p. 40: At the beginning of the twentieth century, Azerbaijan was the center of the oil world, having had its share of Texas-style "gushers." Today it is still an important producer of crude, though it has long fallen from the importance it enjoyed 100 years ago. However, it is not entirely forgotten. The 1,000-mile Baku-Tbilisi-Ceyhan pipeline (the second-longest in the world), as the name implies, links Azerbaijan, Georgia, and Turkey, sends about 1 million barrels per day to the European market, and is of no small geopolitical importance.

p. 45: The ratio between the energy investment in a barrel of Alberta synthetic crude and what goes into producing a barrel of Saudi crude is perhaps best summed up in one of M. King Hubbert's quips: "So long as oil is used as a source of energy, when the energy cost of recovering a barrel of oil becomes greater than the energy content of the oil, production will cease no matter what the monetary price may be" (http://www.hubbertpeak.com/Hubbert/).

CHAPTER 2: **DEMAND SHIFT**

p. 57: To get a glimpse of the extravagance of Ski Dubai, go to http://skidxb.com.

p. 63: The "BRIC" moniker first appeared in the 2003 Goldman Sachs report, "Dreaming With the BRICs" (http://www2.goldmansachs.com/ideas/brics/book/99-dreaming.pdf). The crucial idea was that the economies of Brazil, Russia, India and China will soon overtake those of the more developed countries by their sheer size.

p. 65: For an analysis of how OPEC's own consumption growth is eating into its export capacity, and the attendant implications for world oil supply see Jeff Rubin and Peter Buchanan "OPEC's Growing Call On Itself," September 10, 2007 (http://research.cibcwm.com/res/Eco/EcoResearch. html).

p. 69: The figures for per capita oil consumption around the world are based on data from the CIA World Factbook (https://www.cia.gov/library/publications/the-world-factbook/).

p. 74: Saudi Arabia's foray into growing its food outside its borders is part of a global trend. Libya is leasing land in Ukraine in exchange for oil, South Korea is buying land in Madagascar, China, which already grows food in Australia, is taking over farmland in Cameroon, Laos, Kazakhstan, Burma, Uganda, and the Philippines. The Gulf states have millions of acres in Indonesia, Pakistan, Sudan and Egypt under contract, and Japan has had plenty of prime American agricultural land under lease since the 1980s (http://www.chicagotribune.com/news/nationworld/chi-global-land_bddec14,0,1662931.story).

p. 79: For a thorough analysis of depletion at Ghawar, based on numerous internal reports by geologists and engineers in Saudi Aramco itself, see Matt Simmon's *Twilight in the Desert*, published by Wiley in 2005. Aside from Colin Campbell, Matt Simmons, whom I once shared a podium with many years ago in Calgary, was one of the first people to alert me about the growing dangers of oil depletion. He has been a lonely but very prescient voice for decades now in the US oil and gas industry.

CHAPTER 3: **HEAD FAKES**

p. 87: Daniel Khazzoom and Leonard Brookes are modern day exponents of the Jevons paradox. Both Khazzoom and Brookes independently studied the impact of energy saving measures on energy consumption following the two OPEC oil shocks. Their counter-intuitive finding that gains in energy efficiency boost energy consumption instead of promoting energy conservation is now frequently referred to as the Khazzoom-Brookes postulate. For a more in-depth explanation of the Postulate, see chapter four of George Monbiot's *Heat*, published by Doubleday Canada, 2006.

p. 88: Henry Jevons was one of the most influential economists in Britain during the 19th century and the father of modern day utility theory in economics. He was also one of the first economists to recognize and address the issue of resource depletion, and as such, has attracted more recent attention from those interested in the concept of peak oil. Jevons pioneered the concept of the efficiency paradox in his 1865 treatise "The Coal Question," noting how coal-saving technological change in the production of steel ultimately boosted the demand for coal. The phenomenon was initially known as Jevon's paradox.

p. 91 The fuel efficiency of a 1908 Ford Model T was somewhere between 25 and 17 miles per gallon. Assuming it is the lower figure, many SUVs do worse than their forebears 100 years ago; and if it is the higher figure, even many sleek family vehicles do worse (http://cbs5.com/local/Model.T.Ford.2.434954.html).

p. 95: The story about the threat posed to the British electrical grid by soccer supporters watching their new plasma televisions comes from "Plasma screens threaten eco-crisis," by David Smith and Juliette Jowit, which appeared in *The Observer* August 13, 2006 (http://www.guardian.co.uk/environment/2006/aug/13/energy.nuclear industry).

p. 103: For more information on the environmental impacts of the biofuel boom, see two of George Monbiot's columns in the *Guardian*: "A Lethal Solution" (March 27, 2007) and "An Agricultural Crime Against Humanity" (November 6, 2007). See also "Biofuel Backfiring," by Paul Watson, in the *Los Angeles Times* (October 19, 2008).

The surprising calculation that biofuel can be hundreds of times worse for the environment than fossil fuels comes from a study called "Climate Change and Energy: The True Cost of Biofuels," produced by The Nature Conservancy and the University of Minnesota. It is the carbon released by the destruction of grassland, rainforests, peatlands or savannas in countries like Brazil and the U.S. that adds so much to the carbon footprint of the tailpipe emissions and fossil fuel inputs for the feedstock.

CHAPTER 4: **HEADING FOR THE EXIT LANE**

p. 111: During the breakup of Yugoslavia, the Belgrade government tried to control the exchange rate of the dinar, with predictable results, particularly in a largely black-market economy imposed by UN sanctions. Inflated dinars were of no use to smugglers, since they had to buy gasoline, or drugs, or whatever else it was they were bringing into the country, in some other currency, usually German marks. Interestingly, the mark continued to circulate as unofficial currency in former Yugoslavian republics even after the introduction of the Euro in 1999.

p. 113: Data for US and UK driving habits were derived from "Car Ownership, Travel and Land Use: A Comparison of the US and Great Britain," by G. Guiliano and J. Dargay Working Paper Number 1006, January 11-15, 2005, Transport Studies Unit, Oxford University Centre for the Environment (http://www.tsu.ac.uk/).

p. 114: The rankings of countries by vehicle ownerships comes from http://www.nationmaster.com/graph/tra_mot_veh-transportation-motor-vehicles, and seems to be open to dispute. A January 2009 story in *The Economist* (http://www.economist.com/daily/chart-gallery/displayStory.cfm?story_id=12714391&source=features_box 4) challenges the US's title as the land of the car, but the trend remains the same. With the exception of tiny Luxembourg, it is the wealthy, geographically spread-out nations that have the most cars per capita.

pp. 117-19: To read more on the history of suburban expansion and the death of public transit, see Howard Kunstler's *The Geography of Nowhere*, published by Touchstone Press in 1993, a sustained and unflinching analysis of the cultural and economic costs of engineering a society around the car. It is also worth looking at a website called "Detroit Transit History" (http://www.detroittransithistory.info), compiled by a former Detroit bus driver.

pp. 129-33: The definitive history of GM's EV-1 is told in the documentary *Who Killed the Electric Car?* The story of the Senate Sub-committee on Antitrust is told admirably in a May 2006 article in *The Nation* by Morton Mintz ("Road to Perdition": http://www.thenation.com/doc/20060612/mintz). An assistant sub-committee counsel, Bradford

Snell, tells the story as well in an essay called "The Streetcar Conspiracy: How General Motors Deliberately Destroyed Public Transit" (http://saveourwetlands.org/ streetcar.htm).

p. 133: A stripped-down, chain-driven, chopped version of the everyday Opel T-1 earned a place in the Guinness Book of Records in 1973 by wringing out 377 mpg at a steady 30 miles per hour. While hardly as sleek as a Prius, or even as practical, the fact that a pretty much off-the-rack engine from the days before many people had televisions could beat today's models by more than a couple of hundred mpg shows just how little progress we've made (http://seattlepi.nwsource. com/local/351903_needle20.html).

p. 135: A barrel of oil contains 6.1 GJ (5.8 million Btu), equivalent to 1,700 kWh. The average monthly household electricity consumption in the US is 936 kWh (http://tonto.eia.doe.gov/ask/electricity_ faqs.asp), or 11,232 kWh per year. So the 13 million barrels of oil used by the American vehicle fleet would generate 221 billion kWh, or enough to keep the lights on in 1,967,595 homes for a year.

p. 137: The Finnish Olikiluoto 3 project, built by the French company Areva, was to be the first nuclear reactor built in Europe in thirty years, and the flagship of the so-called third generation reactors. But despite the bullishness, construction has been hampered by cost over-runs and safety concerns, and governments around the world look on with concern as problems mount (http://www.guardian.co.uk/business/ 2009/jan/14/areva-nuclear-finland-olkiluoto).

p. 137-38: The trouble with renewable energy is that the wind and sun are intermittent, and can't be relied on to keep the electricity flowing when we want it. Solar panels aren't much good on a winter night in Canada. Meanwhile, the electrical grid requires a constant flow. For this reason, even if we had all the solar arrays and wind farms we could build, we would still need to keep other generators on hand (coal, gas, nuclear, or hydro) to supply base load. However, a study commissioned by the German Ministry of the Environment, Nature Conservancy and Nuclear Safety shows that it is possible to meet electrical demand largely with renewables, in large part by getting your electricity from places where the sun *does* shine nearly all the time. In other words, Europe could get much of its power from North Africa

(http://www.dlr.de/tt/Portaldata/41/Resources/dokumente/institut/
system/projects/TRANS-CSP_Full_Report_Final.pdf). Presumably,
Canada and the northern US states could also harness solar energy
from Mexico.

p. 138-39: CCS involves stripping the carbon dioxide from exhaust,
then pumping it underground into spent oil or gas wells, or deep
enough underwater that it won't circulate or rise to the surface. The
idea of taking the greenhouse gases out of our fossil fuel consumption certainly has its appeal, and a lot of hope and money is being
invested in the technology, but it is not a way to continue with business as usual. In addition to the difficulties posed by the sheer scale
of the infrastructure required and the fact that the technology is
untested at anything like the scale required to make a difference,
there is the fact that the process is energy-intensive and would
increase a plant's fuel requirements by as much as 40 percent,
taking back the efficiency gains of the past few decades and raising
costs by as much as 90 percent (http://www.ipcc.ch/pdf/special-
reports/srccs/srccs_wholereport.pdf).

p. 139-40: The most common way of producing hydrogen is by processing natural gas, which makes the fuel a pretty inappropriate
response to the problems of both resource depletion and climate
change. See "The Car of the Perpetual Future," by Phil Wriggle-
sworth, in the September 4, 2008, edition of *The Economist*.
(http://www.economist.com/science/tq/displaystory.cfm?story_id=
11999229).

CHAPTER 5: **COMING HOME**

pp. 146-153: Transoceanic shipping cost data and tariff equivalent calcu-
lations taken from Jeff Rubin and Benjamin Tal "Will Triple Digit
Oil Prices Reverse Globalization?," May 27, 2008 (http://research.
cibcwm.com/res/Eco/Research./html).

p. 152: The account of Crown Battery moving production from Mexico
back to the US comes from a June 13, 2008, story in *The Wall Street
Journal* by Timothy Aeppel ("Stung by Soaring Transport Costs,

Factories Bring Jobs Home Again": http://www.uawregion1a.org/
News%20Articles/Newpaper/Det.%20News/(Microsoft%20Word%20-
%20Stung%20by%20soaring%20Transport%20Costs,%20Factories%
20Bring%20Jobs.pdf).

CHAPTER 6: **THE OTHER PROBLEM WITH FOSSIL FUELS**

p. 157: The issue of climate change and the question of viable targets is
dealt with pretty cursorily in this chapter, but it certainly bears
further reading. James Hanson's argument that current targets of 550
or even 450 parts per million of CO2 are far too high appears in
"Target Atmospheric CO2: Where Should Humanity Aim?"
(http://arxiv.org/ pdf/0804.1126v3). Hanson, head of NASA's
Goddard Institute for Space Studies in New York, has been perhaps
the most urgent voice warning of the dangers of climate change over
the past decades.

p. 159: The chapter's figures on the pace of global greenhouse gas emissions come from The Global Carbon Project (www.globalcarbonpro-
ject.org/global/pdf/GCP_CarbonBudget_2007.pdf).

p. 160: The figures for French nuclear generating capacity come from
the World Nuclear Association (http://www.world-nuclear.org/info/
inf40.html).

p. 163: The figures for British, Canadian and Australian reliance on coal
for electricity generation come from the Energy Information
Administration's "Country Analysis Briefs" (http://www.eia.doe.gov/
emeu/cabs/index.html).

p. 164: The projections for coal consumption in Taiwan, Vietnam,
Indonesia and Malaysia come from the Energy Information
Administration's "International Energy Outlook 2008" (http://www.eia.
doe.gov/oiaf/ieo/coal.html).

p. 165: It is hard to exaggerate the importance of glacial runoff not
only for China but also for millions of people in Southeast Asia,
India and Pakistan. Once the glaciers are gone, their source of fresh
water will be too. (http://www.sciencedaily.com/releases/2007/04/
070410134724.htm).

pp. 166-172: For a somewhat technical but thorough analysis of the contribution made by China's export sector to the country's massive carbon emissions see Weber et al. "The Contribution of Chinese Exports to Climate Change" in *Energy Policy* (2008), Doi 20.1016/j.enpol2008.06.009.

CHAPTER 8: GOING LOCAL

p. 209: For a history of coffee and its integral role in history, see *A History of the World in Six Glasses*, by Tom Standage, published by Doubleday Canada, Toronto, 2005, pp. 133-172.

p. 218: For a discussion of the demise of the British apple, see George Monbiot's "Fallen Fruit" (http://www.monbiot.com/archives/2004/10/30/fallen-fruit/). An interesting thumbnail of the distance traveled by items in a British supermarket appears in the *Guardian* ("Miles and Miles and Miles": http://www.guardian.co.uk/lifeand-style/2003/may/10/foodanddrink.shopping6).

p. 219 The transportation figures for Australia come from "Food Miles in Australia" (http://www.ceres.org.au/projects/CERES_Report_%20 Food_Miles_in_Australia_March08.pdf).

p. 220: The account of grow-ops sprouting in abandoned suburbs comes from a February 7, 2009, story by Damien Cave in the *New York Times* ("In Florida, Despair and Foreclosures," http://www.nytimes.com/2009/02/08/us/08lehigh.html).

p. 223: The issue of energy dependency of agriculture has been explored in great depth in many books and articles. *Against the Grain*, by Richard Manning, published by North Point Press in 2004, is a fascinating and unsettling history of farming and agribusiness. The comparison of agricultural energy inputs to nuclear bombs comes from the same author's essay "The Oil We Eat" in the February 2004 issue of *Harper's* (http://harpers.org/archive/2004/02/0079915).

p. 225: The quotation from World Bank president Robert Zoellick comes from an October 8, 2008, story by Heather Stewart in the *Guardian* (http://www.guardian.co.uk/business/2008/oct/08/world-bank.food).

p. 229: The figures for the energy intensity of a DRAM chip come from *Energy in Nature and Society*, by Vaclav Smil, published by MIT Press in 2008.

p. 231: The fuel consumption data for a Boeing 767 comes from a February 22, 2008, article in *Salon*: "Ask the Pilot," by Patrick Smith (http://www.salon.com/tech/col/smith/2008/02/22/askthepilot265/print.html). If the average American drives 12,000 miles a year in a mid-sized car getting 26.7 mpg (as per Project America: http://www.project.org/info.php?recordID=384), that works out to about 450 gallons. So the return flight is about three months worth of driving.

p. 233: The story from *The Times* about Ryanair's fuel policy (Steven Swinford, "Ryanair fuel ration angers pilots") appeared in the August 31 edition and can be seen at http://business.timesonline.co.uk/tol/business/industry_sectors/transport/article4641399.ece.

p. 236: The figures for global tourism come from the World Travel and Tourism Council (http://www.wttc.org/eng/Tourism_News/Press_Releases/Press_Releases_2008/World_Citizens_-_a_global_partnership_for_Travel_and_Tourism/) and an August 31, 2008, article in the *Washington Post* by Elizabeth Becker ("Don't Go There": http://www.washingtonpost.com/wp-dyn/content/article/2008/08/29/AR2008082902337.html?hpid=opinionsbox1).

p. 238: The figures for the consequences of oil prices for sub-Saharan Africa comes from an IEA report entitled "Analysis of the Impact of High Oil Prices on the Global Economy (http://www.iea.org/Textbase/Papers/2004/High_Oil_Prices.pdf).

p. 240: Figures for percentages of GDP derived from agriculture come from the World Resources Institute (http://earthtrends.wri.org).

p. 241: Figures for the percentage of GDP in the Philippines derived from overseas remittances comes from the Global Strategy Institute (http://forums.csis.org/gsionline/?p=565).

p. 242: The organization calling itself America's Leadership Team for Long Range Population-Immigration-Resource Planning is an umbrella group composed of the American Immigration Control Foundation, Californians for Population Stabilization, the Federation for American Immigration Reform and NumbersUSA. One of their advertisements makes the case that "America has problems—huge problems," then goes on to say "all of these problems are caused by a large and fast-growing population."

p. 247: The figure of $50 per ton to reduce emissions comes from a report by McKinsey & Company called "Reducing US Greenhouse Gas Emissions: How Much and at What Cost?" (http://www.mckinsey.com/clientservice/ccsi/greenhousegas.asp).